Forensic Chemistry of Substance Misuse
A Guide to Drug Control

Forensic Chemistry of Substance Misuse
A Guide to Drug Control

L. A. King

RSCPublishing

ISBN: 978-0-85404-178-7

A catalogue record for this book is available from the British Library

© L. A. King 2009

All rights reserved

Apart from fair dealing for the purposes of research for non-commercial purposes or for private study, criticism or review, as permitted under the Copyright, Designs and Patents Act 1988 and the Copyright and Related Rights Regulations 2003, this publication may not be reproduced, stored or transmitted, in any form or by any means, without the prior permission in writing of The Royal Society of Chemistry or the copyright owner, or in the case of reproduction in accordance with the terms of licences issued by the Copyright Licensing Agency in the UK, or in accordance with the terms of the licences issued by the appropriate Reproduction Rights Organization outside the UK. Enquiries concerning reproduction outside the terms stated here should be sent to The Royal Society of Chemistry at the address printed on this page.

Published by The Royal Society of Chemistry,
Thomas Graham House, Science Park, Milton Road,
Cambridge CB4 0WF, UK

Registered Charity Number 207890

For further information see our web site at www.rsc.org

Preface

"Gold is worse poison to a man's soul, doing more murders in this loathsome world, than any mortal drug"

William Shakespeare (Romeo and Juliet)

An earlier publication[1] described the UK drugs legislation from the viewpoint of a forensic scientist. In the current book, an opportunity has been taken to rearrange and expand the material and improve clarity, to include the changes that have occurred in the past six years, and, more importantly, to widen the scope and the intended audience. Given the high political profile of drug misuse and the large number of offenders regularly prosecuted, drugs legislation is subjected to a high level of scrutiny by the Courts. The legislation is also technically complex with areas that are rarely explored. It follows that there is need for all participants in the legal process to have some familiarity with the underlying chemical principles. This book is intended to provide that background understanding and complements other publications that deal primarily with legal interpretation and the case law that has built up over the past three decades.

This book is largely based on UK law and practice, and provides a description of the current legislation. However, this is placed into the context of the United Nations drug control conventions, and, where appropriate, compared with the US legislation. Sitting between national

---

[1] L.A. King, *The Misuse of Drugs Act: A Guide for Forensic Scientists* – see Bibliography

legislations and the international drug control treaties, the European Union (EU) has a supranational role. The EU is having an increasing impact on domestic law. Thus, apart from precursor legislation, which derives from the United Nations Convention Against Illicit Traffic in Narcotic Drugs and Psychotropic Substances 1988, the EU has specific competence in the area of "new psychoactive substances", formerly known as "new synthetic drugs".

While most countries have chosen to implement only the essential elements required in international law by the 1961 and 1971 United Nations Conventions, a few have extended the scope to a wider range of substances. Examples of other countries' approaches to drug control in regard to generic/analogue controls and emergency legislation are provided. The structure-specific generic controls in the Misuse of Drugs Act are comprehensively covered with more examples. Yet other generic controls derive from the international drug control treaties, and are therefore common to the law of many countries. An unusual feature of UK drugs law, shared with that of the US and only a few other countries, is that it includes a large number of anabolic steroids. These substances not only lack psychoactivity, but even include testosterone: a steroid that occurs naturally in human and other mammalian tissues.

Although many new substances have been brought under control in recent years, the list of potential candidates has increased even more. Throughout the 1990s most "new synthetic drugs" were either ring-substituted phenethylamines or, less commonly, substituted tryptamines. In the last six years, clandestine drug manufacturers seem to have largely exhausted this chemical repertoire and have now diversified into a much more heterogeneous group of substances. Nevertheless, these psychoactive novelties continue to be mostly CNS stimulants or compounds with a pharmacology having some resemblance to that of the well-established drug MDMA (3,4-methylenedioxymethylamphetamine; ecstasy).

As a subtheme to the arguments about relative harmfulness, considerable time and energy have been expended in the UK, particularly since 2002, on the specific classification of cannabis, and to a lesser extent of certain other substances. The irony is that, despite three major reviews and an intermediate period when its status was changed, cannabis will soon be back to where it was in late 2003 and had been since 1971. When it is recognised that few substances have been reclassified since 1971, many observers might conclude that the system is largely impervious to change and should be replaced for that reason alone.

All States have to address the question of whether certain activities with certain substances should attract heavier penalties than others, regardless of whether those penalties are defined in the criminal law or are civil penalties or merely administrative sanctions. In recent years, many critical questions have been raised as to whether, after nearly 40 years, the UK drugs legislation is still fit for purpose. To a large extent such questions have centred on the classification of substances and their relative harmfulness. These concerns are relevant to all legislations, since it is a general principle of drug laws that there should be a correlation between harm, either to the individual or society, and the penalties associated with various offences.

As part of the critical light that now shines on drug control, not only are concerns being raised about the substances that are scheduled, but anomalies with society's approach to nonscheduled substances are becoming clearer. The most obvious of these are alcohol and tobacco, which together cause far more damage to society and to individuals than all of the scheduled substances combined. Yet these substances are often not even regarded as drugs. But these "socially acceptable" substances are by no means free of controls. The law determines such matters as who may sell them, where, when and to whom. Furthermore, the social acceptance of alcohol has a strong cultural and religious link. If we take a strict line about relating legal control to harmfulness and by relating harmfulness largely to the pharmacological and toxicological properties of those substances then we must recognise that the social drugs lie on a continuum of harm with all other substances and do not belong in some different dimension. From here it is a short step to examine our attitude to all harmful substances and ask how they should fit into the scale. Such substances include simple poisons, drug and weapon precursors, industrial solvents, established medicines, harmful materials in the workplace, the social drugs and those, often innocuous, substances intended for use as cutting agents.

This book is aimed not only at forensic scientists but also at police and customs officers, lawyers and all those with an interest in drugs legislation. There is coverage of the many problem areas that arise in the forensic interpretation of analytical results. For chemists, the extensive use of molecular structures in the text allows a complete and easier comprehension of the chemical background to the legislation. It is not a guide to general aspects of the law, stated cases, sentencing policy or related legislation although some of these are dealt with briefly in the Appendices. Also excluded is any comprehensive discussion of chemical analysis, but brief analytical properties of the major drugs are provided, and specific problem areas are mentioned where they have a bearing on interpretation. No account is provided of the wider social dimension to drug abuse, to epidemiology, pharmacology or toxicology, but the interested reader is directed to the Bibliography. Selected references to specific articles and research publications are included in the text as footnotes, but these are not meant to be exhaustive. There is no discussion of the arguments for or against legalisation or decriminalisation of some or all drugs, or to what extent drug misuse is a health problem as opposed to a law-enforcement issue. Finally, it is beyond the scope of this book to provide any recommendations on the presentation of evidence in Court or how analytical results should be set out in reports and statements.

I particularly wish to thank Professor Geoffrey Phillips, Dr John Ramsey and Professor Les Iversen for supporting the concept of this book in early discussions with the Royal Society of Chemistry. Professor Geoffrey Phillips, Rudi Fortson and Ric Treble kindly reviewed a draft manuscript and offered valuable comments.

Leslie A. King,
Hampshire

Contents

Glossary		xix
Chapter 1	**Introduction**	1
	1.1 Drug Misuse	1
	1.2 Abbreviations	3
Chapter 2	**Control of Chemical Substances**	6
	2.1 Introduction	6
	2.2 Poisons	7
	2.3 Pharmaceutical Ingredients and Medicinal Products	8
	2.4 Drug Precursors	10
	2.5 Chemical Weapons and their Precursors	11
	2.6 Solvents and Gases	11
	2.6.1 Volatile Solvents and Gases	11
	2.7 The "Social" Drugs	12
	2.7.1 Alcohol	12
	2.7.2 Tobacco	13
	2.7.3 Caffeine	14
	2.7.4 Khat	15
	2.8 Dangerous Substances	16
Chapter 3	**Nomenclature**	17
	3.1 British Approved Names and International Nonproprietary Names	17

Forensic Chemistry of Substance Misuse: A Guide to Drug Control
By L.A. King
© L.A. King 2009
Published by the Royal Society of Chemistry, www.rsc.org

3.2	Synonyms and Common Terms	18
3.3	Redundancy	20
3.4	The Meaning of "Derivative" in the Misuse of Drugs Act	20
3.5	Dialkyl Derivatives	21
3.6	The Meaning of "Structurally Derived From"	22
3.7	Homologues	22
3.8	"Phenethylamines", "Phenylethylamines", Phenylalkylamines and "Amphetamines"	23

Chapter 4 Drug Control at International and European Level — 24

4.1	United Nations Single Convention on Narcotic Drugs (1961)	24
4.2	United Nations Convention on Psychotropic Substances (1971)	25
4.3	United Nations Convention Against Illicit Traffic in Narcotic Drugs and Psychotropic Substances (1988)	26
4.4	European Initiatives	27
	4.4.1 The Period 1997 to 2005	27
	4.4.2 Developments since 2005	28

Chapter 5 Drug Legislation in the UK — 30

5.1	Historical Background	30
5.2	The Misuse of Drugs Act 1971	31
	5.2.1 The Misuse of Drugs Act – Substances Removed or Reinstated	33
	5.2.2 The "PIHKAL" List	34
	5.2.3 Substances Added to the UN Conventions (1961, 1971) and/or the Misuse of Drugs Act since 2002	35
	5.2.3.1 Amineptine	36
	5.2.3.2 Anabolic Steroids (Derivatives of Androstene)	37
	5.2.3.3 4-Bromo-2,5-dimethoxyphenethylamine (2C-B)	37
	5.2.3.4 Dihydroetorphine	37
	5.2.3.5 γ-Hydroxybutyrate (GHB)	38
	5.2.3.6 Ketamine	39
	5.2.3.7 α-Methyl-4-(methylthio)-phenethylamine (4-MTA)	40

		5.2.3.8	Remifentanil	40
		5.2.3.9	Zolpidem	41
	5.3	The Misuse of Drugs Act: Changes Pending		41
	5.4	The Misuse of Drugs Regulations, 2001		41
	5.5	Drugs Act 2005		44

Chapter 6 Generic Controls in the UK — **45**

    6.1 "Designer drugs" — 45
    6.2 Salts — 48
    6.3 Esters and/or Ethers — 49
        6.3.1 Esters — 50
        6.3.2 Ethers — 51
    6.4 Stereoisomerism — 53
    6.5 Anabolic Steroids — 55
    6.6 Barbiturates — 57
    6.7 Cannabinols — 59
    6.8 Ecgonine Derivatives — 61
    6.9 Fentanyls — 63
    6.10 Lysergide and Derivatives of Lysergamide — 65
    6.11 Pentavalent Derivatives of Morphine — 66
    6.12 Pethidines — 66
    6.13 Phenethylamines — 68
    6.14 Tryptamines — 72
        6.14.1 The "TIHKAL" Drugs — 74

Chapter 7 Natural Products – Problem Areas — **76**

    7.1 Cannabis and Cannabis Resin — 76
        7.1.1 Introduction — 76
        7.1.2 Definitions of Cannabis — 77
        7.1.3 Cannabis Seeds — 78
        7.1.4 Hash oil — 78
        7.1.5 Cannabis-Based Medicines and Dronabinol — 79
        7.1.6 "High-Potency" Cannabis — 80
    7.2 Opium — 80
        7.2.1 Definitions of Opium — 81
        7.2.2 Identification of Opium — 82
        7.2.3 A New Definition of Opium? — 82
    7.3 Poppy-Straw and Concentrate of Poppy-Straw — 82
    7.4 "Magic" Mushrooms — 83
    7.5 Coca Tea — 84

Chapter 8	**Other Problems of Chemical/Legal Interpretation**	**85**
	8.1 Crack Cocaine	85
	8.2 Diagnostic Kits	86
	8.3 Isotopic Variants	86
	8.3.1 A Case History	87
	8.3.2 Implications for the Misuse of Drugs Act	88
	8.4 Low-Dosage Preparations	88
	8.5 Medicinal Products	89
	8.6 Purity, Potency and Drug Content	91
	8.7 Preparations Designed for Administration by Injection	92
	8.8 Supply of Meat Products	93
Chapter 9	**Candidates for Future Control**	**94**
	9.1 New Psychoactive Substances in the "Post-PIHKAL" Era	94
	9.2 Other Phenylalkylamines	96
	9.2.1 N-Substituted Phenethylamines	96
	9.2.2 Other Side-Chain Phenylalkylamine Variants	97
	9.2.3 Other Ring-Substituted Phenethylamines	99
	9.3 1-Benzylpiperazine and other Derivatives of Piperazine	99
	9.3.1 N-Substituted Piperazines: A Possible Generic Definition	100
	9.4 Substituted Cathinones	102
	9.5 Substituted Indans, Indenes and Tetralins	105
	9.6 γ-Butyrolactone (GBL), 1,4-Butane-diol (1,4-BD) and Related Substances	107
	9.7 Ephedrine and Pseudoephedrine	107
	9.8 Additional Anabolic Steroids	108
	9.9 Substances under Review by WHO	108
	9.9.1 Zopiclone	109
	9.9.2 Tramadol	110
	9.9.3 Butorphanol	110
	9.9.4 Oripavine	111
	9.10 Diphenyl-2-pyrrolidinylmethanol	111
	9.11 Methylhexaneamine	112
	9.12 Modafinil	112
	9.13 Phenazepam	113
	9.14 Miscellaneous Opioids	113

	9.15	Cognitive Enhancers	114
	9.16	Miscellaneous Natural Products Containing Psychoactive Drugs	115
		9.16.1 Peyote and Other Cacti	115
		9.16.2 "Morning Glory" Seeds	116
		9.16.3 Plants Containing Tryptamines	117
	9.17	Carisprodol	117
	9.18	Miscellaneous Hypnotics and Other Substances	118
	9.19	Cutting Agents and Adulterants	118

Chapter 10 Generic and Analogue Control – International Comparisons   **119**

	10.1	Generic Definitions in New Zealand	119
		10.1.1 Amphetamine Derivatives	120
		10.1.2 Methaqualone Derivatives	121
		10.1.3 Tryptamine Derivatives	122
	10.2	Drug "Analogues"	123
		10.2.1 US Analogue Control	123
		10.2.2 New Zealand Analogue Control	124
	10.3	Emergency Scheduling Provisions	125

Chapter 11 The Drug Classification Debate   **126**

	11.1	Introduction	126
	11.2	The 1979 Review by ACMD	127
	11.3	The Independent Enquiry into the Misuse of Drugs Act (2000)	127
	11.4	Reclassification of Cannabis (2001–4)	128
	11.5	Home Affairs Select Committee (2001–2)	129
	11.6	Review of Cannabis (2005–6)	130
	11.7	Home Office Proposals for Reviewing the Classification System (2006)	130
	11.8	Methylamphetamine (2006)	130
	11.9	Select Committee on Science and Technology (2006)	132
	11.10	Royal Society of Arts Report (2007)	132
	11.11	Scale of Drug Harm (2007)	133
	11.12	Review of Cannabis (2007–8)	135
	11.13	Review of Ecstasy (2008)	136
	11.14	Systematic Review of other Substances (2009 Onwards)	137
	11.15	The "Precautionary Principle"	137
	11.16	Conclusions	138

Chapter 12	**The Future of "Substance" Legislation in the UK**		**141**
	12.1 Introduction		141
	12.2 Offence-Dependent Classification		142
		12.2.1 The Division of Offences into Two Groups	142
		12.2.2 The Proposed Class D	145
		12.2.3 The Overall Scale of Harm	145
		12.2.4 Generically Defined Substances	146
		12.2.5 Precursor Chemicals	146
		12.2.6 Existing Controlled Drugs	147
		12.2.7 Other Chemicals	147

General Bibliography   148

Appendix 1 Modification and Amendment Orders to the Misuse of Drugs Act 1971   151

A1.1	The Misuse of Drugs Act 1971 (Modification) Order 1973 (S.I. 771)	151
A1.2	The Misuse of Drugs Act 1971 (Modification) Order 1975 (S.I. 421)	151
A1.3	The Misuse of Drugs Act 1971 (Modification) Order 1977 (S.I. 1243)	151
A1.4	The Misuse of Drugs Act 1971 (Modification) Order 1979 (S.I. 299)	151
A1.5	The Misuse of Drugs Act 1971 (Modification) Order 1983 (S.I. 765)	152
A1.6	The Misuse of Drugs Act 1971 (Modification) Order 1984 (S.I. 859)	152
A1.7	The Misuse of Drugs Act 1971 (Modification) Order 1985 (S.I. 1995)	152
A1.8	The Misuse of Drugs Act 1971 (Modification) Order 1986 (S.I. 2230)	152
A1.9	The Misuse of Drugs Act 1971 (Modification) Order 1989 (S.I. 1340)	152
A1.10	The Misuse of Drugs Act 1971 (Modification) Order 1990 (S.I. 2589)	152
A1.11	The Misuse of Drugs Act 1971 (Modification) Order 1995 (S.I. 1966)	153
A1.12	The Misuse of Drugs Act 1971 (Modification) Order 1996 (S.I. 1300)	153
A1.13	The Misuse of Drugs Act 1971 (Modification) Order 1998 (S.I. 750)	153

A1.14	The Misuse of Drugs Act 1971 (Modification) Order 2001 (S.I. 3932)	153
A1.15	The Misuse of Drugs Act 1971 (Modification) Order 2003 (S.I. 1243)	153
A1.16	The Misuse of Drugs Act 1971 (Modification) (No. 2) Order 2003 (S.I. 3201)	153
A1.17	The Misuse of Drugs Act 1971 (Amendment) Order 2005 (S.I. 3178)	154
A1.18	The Misuse of Drugs Act 1971 (Amendment) Order 2006 (S.I. 3331)	154

Appendix 2 The Misuse of Drugs Regulations (Schedule 4)   **155**

Appendix 3 The Misuse of Drugs Regulations (Schedule 5)   **158**

Appendix 4 Drug "Intermediates" in the Misuse of Drugs Act 1971   **160**

Appendix 5 Drug Precursors   **161**

Appendix 6 A Brief History of the Legal Status of Hash Oil   **164**

Appendix 7 Other Drug-Related Legislation   **166**

A7.1	Road Traffic Act 1972	166
A7.2	The Customs and Excise Management Act 1979	166
A7.3	The Drug Trafficking Act 1994	166
A7.4	The Crime and Disorder Act 1998	167
A7.5	The Criminal Justice and Police Act 2001	167
A7.6	The Criminal Justice Act 2003	167
A7.7	The Criminal Justice (International Cooperation) Act 1990; Controlled Drugs (Drug Precursors) (Intra-Community Trade) Regulations 2008; Controlled Drugs (Drug Precursors) (Community External Trade) Regulations 2008	167

Appendix 8 Relevant Stated Cases   **168**

A8.1	Usability	168
A8.2	Generic Legislation	169
A8.3	Cannabis and Cannabis Resin	169

A8.4	Crack Cocaine	169
A8.5	Salts and Stereoisomers	169

Appendix 9   Sentencing Guidelines — **170**

Appendix 10   Profiles of the Major Drugs of Misuse — **173**

A10.1	Amphetamine		173
	A10.1.1	Introduction	173
	A10.1.2	Chemistry	174
	A10.1.3	Physical form	174
	A10.1.4	Pharmacology	174
	A10.1.5	Synthesis and Precursors	175
	A10.1.6	Mode of Use	176
	A10.1.7	Other Names	176
	A10.1.8	Analysis	176
	A10.1.9	Control Status	176
	A10.1.10	Medical Use	177
A10.2	Cannabis		177
	A10.2.1	Introduction	177
	A10.2.2	Chemistry	178
	A10.2.3	Physical Form	178
	A10.2.4	Pharmacology	178
	A10.2.5	Origin	179
	A10.2.6	Mode of Use	179
	A10.2.7	Other Names	180
	A10.2.8	Analysis	180
	A10.2.9	Control Status	180
	A10.2.10	Medical Use	181
A10.3	Cocaine and Crack		181
	A10.3.1	Introduction	181
	A10.3.2	Chemistry	181
	A10.3.3	Physical Form	182
	A10.3.4	Pharmacology	182
	A10.3.5	Origin/Extraction	182
	A10.3.6	Mode of Use	183
	A10.3.7	Other Names	183
	A10.3.8	Analysis	183
	A10.3.9	Control Status	184
	A10.3.10	Medical Use	184
A10.4	Diamorphine (Heroin)		184
	A10.4.1	Introduction	184
	A10.4.2	Chemistry	185
	A10.4.3	Physical Form	185
	A10.4.4	Pharmacology	185

		A10.4.5	Origin/Extraction	186
		A10.4.6	Mode of Use	186
		A10.4.7	Other Names	187
		A10.4.8	Analysis	187
		A10.4.9	Control Status	187
		A10.4.10	Medical Use	187
	A10.5	Lysergide (LSD)		187
		A10.5.1	Introduction	188
		A10.5.2	Chemistry	188
		A10.5.3	Physical Form	188
		A10.5.4	Pharmacology	188
		A10.5.5	Synthesis and Precursors	189
		A10.5.6	Mode of Use	190
		A10.5.7	Other Names	190
		A10.5.8	Analysis	190
		A10.5.9	Control Status	191
		A10.5.10	Medical Use	191
	A10.6	MDMA		191
		A10.6.1	Introduction	191
		A10.6.2	Chemistry	192
		A10.6.3	Physical Form	192
		A10.6.4	Pharmacology	192
		A10.6.5	Synthesis and Precursors	193
		A10.6.6	Mode of Use	193
		A10.6.7	Other Names	193
		A10.6.8	Analysis	194
		A10.6.9	Control Status	194
		A10.6.10	Medical Use	194
	A10.7	Methylamphetamine		194
		A10.7.1	Introduction	194
		A10.7.2	Chemistry	195
		A10.7.3	Physical Form	195
		A10.7.4	Pharmacology	195
		A10.7.5	Synthesis and Precursors	196
		A10.7.6	Mode of Use	197
		A10.7.7	Other Names	197
		A10.7.8	Analysis	197
		A10.7.9	Control Status	198
		A10.7.10	Medical Use	198
Appendix 11	Field Tests and the "Guilty Plea Policy"			**199**
Appendix 12	Purities and Drug Content of Illicit Substances			**201**
Appendix 13	Prices and Wrap Sizes of Illicit Drugs			**203**

Appendix 14	Useful Websites	204
Appendix 15	The Misuse of Drugs Act – Schedule 2 (Parts I to III)	205
Appendix 16	The Misuse of Drugs Act 1971 – Schedule 2 (Part IV)	215
Appendix 17	Phenethylamines added to the Misuse of Drugs Act in 2001	217
Appendix 18	Structural Classification of the Phenethylamines added to the Misuse of Drugs Act in 2001	221
Appendix 19	Molecular Structures of the Phenethylamines added to the Misuse of Drugs Act	222
Appendix 20	Derivatives of Tryptamine	229
Subject Index		233

Glossary

Terms emboldened are themselves defined.

Addiction:
For most purposes, addiction is synonymous with **dependence**.

Adulterant:
Often synonymous with **cutting agent**.

Agonist:
A drug that mimics the effect of **neurotransmitters** or other endogenous molecules. It has the opposite effect to an antagonist.

Alkali:
Usually an inorganic base such as sodium hydroxide, sodium carbonate/bicarbonate. By combining with the chemically bound acid residue, an alkali is used, for example, to convert a salt into the free **base**.

Alkaloid:
A naturally occurring nitrogenous **base**.

Aluminium foil method:
A type of **reductive amination** that requires little equipment. A precursor ketone (*e.g.* **P2P, PMK**) is reacted in ethanol with aluminium metal pieces, an **amine** and a mercuric chloride catalyst. When the amine is methylamine and the ketone is PMK then the product is MDMA.

Amine:
A chemical group comprising a nitrogen atom attached to one or more carbon atoms and one or more hydrogen atoms. Amines are typically **bases**.

Analgesic:
A substance that reduces the sensation of pain. See also **narcotic analgesic**.

Analogue control:
The inclusion in legislation of a definition that covers a family of substances. It is less specific than **generic control**, and may be based on the concept of "similarity in chemical structure" as well as "similarity in pharmacological activity" to the parent substance.

Anti-tussive:
A substance that reduces the cough reflex.

Base:
A nitrogenous substance, sometimes known as an **alkaloid** when derived from plant material, which reacts with acids to form a **salt**. Many bases are insoluble in water but soluble in organic solvents.

Cannabinoid:
One of a group of compounds found only in *Cannabis sativa* including cannabidiol (**CBD**), cannabinol (**CBN**) and tetrahydrocannabinol (**THC**).

CAS:
Chemical Abstracts Service. A body responsible for indexing the world's chemistry-related literature and patents.

CBD:
Cannabidiol (CBD) is one of several cannabinoids in *Cannabis sativa*. It has anti-psychotic effects and occurs at a higher concentration in cannabis resin than in herbal cannabis.

CBN:
Cannabinol (CBN) is a **cannabinoid** and an oxidation product of **THC**. It is normally only found in aged samples of cannabis and cannabis resin.

Cutting agent:
A substance added as a diluent to a drug. It may be inert or pharmacologically active. Such diluents can be found in illicit powders as well as tablets where the term might also include tablet binders.

Decriminalisation:
The removal of a conduct or activity from the criminal law, but with a remaining prohibition that may be dealt with by civil or administrative methods.

Glossary

Dependence:
Drug dependence is a process whereby repeated use leads to increasing difficulty in stopping. It is a complex phenomenon whose nature differs from drug to drug, but which is also dependent on the duration and quantity that is used as well as characteristics of the user. Dependence is also related to the pleasure that a drug gives: the more immediate pleasure a user experiences, the more likely it is to cause dependence. It is reflected in an increasing reliance on the drug and by symptoms of withdrawal when users reduce their consumption or attempt to stop. The term dependence is used by the World Health Organisation (WHO) in preference to **addiction**.

Diastereoisomers:
When a molecule contains two centres of asymmetry, it can form four diastereoisomers (*i.e.* two pairs of **enantiomers**). Thus, ephedrine, for example, exists as four diastereoisomers, two of which are known as pseudoephedrine.

Dopamine:
An example of a **neurotransmitter**, it is a naturally occurring substituted **phenethylamine**. Substances that interact with the dopamine receptor are said to be dopaminergic.

Drug Abuse:
The use of a pharmacologically active substance for nonmedical purposes.

Drug Misuse:
For most purposes, drug misuse is synonymous with **drug abuse**.

Ecstasy:
Originally used to describe MDMA, but since generalised to describe a wide range of substituted **phenethylamines** and, less precisely, certain unrelated substances.

Empathogen:
A substance that produces empathy with others, most often applied to MDMA and related drugs. See also **entactogen**.

Enantiomer:
One of a pair of **stereoisomers** arising from the presence in a molecule of an asymmetric carbon atom. The two enantiomers are mirror images of each other, with left- and right-handed forms denoted S (sinister) and R (rectus), respectively. Enantiomeric pairs were previously denoted by terms such as d and l or by the symbols $(+)$ and $(-)$.

Endogenous anabolic steroid:
One that is naturally produced by the (human) body.

Entactogen:
A substance that produces a socialising effect and desire for contact, most often applied to MDMA and related drugs. See also **empathogen**.

Exogenous anabolic steroid:
One that is not naturally produced by the (human) body.

Generic control:
The inclusion in legislation of a definition that covers a family of substances. At one level this includes esters or ethers of a parent molecule – an example that derives from the 1961 United Nations Single Convention on Narcotic Substances. More elaborate generic definitions are based on substitution patterns in a parent molecule where the type, number and position of substituents may be precisely specified. A consequence of generic control is that it may subsume substances with varied pharmacological activity or even none at all. The generic approach should be contrasted with **analogue control**.

Half-life:
The time required for the concentration of a drug in a tissue (*e.g.* blood) to fall to 50% of its initial value.

Hallucinogen:
A substance that produces, as a main effect, perceptual distortions, especially visual and auditory. The effects can also extend beyond perceptions to changes of thought, mood and personality integration (self-awareness). The term is somewhat misleading as some so-called hallucinogenic substances do not cause true hallucinations (*i.e.* sensory perceptions in the absence of external stimuli). However, the term is widely accepted by the scientific community. Hallucinogen is often now used for substances that were once described as psychedelic.

Homologue:
When a series of chemical compounds differ only by a constant structural element, they are said to form a homologous series. Thus, MDA, MDMA, MDEA are homologues where each successive member differs from the previous structure by a methylene moiety (CH_2).

Hydrate:
Some **salts** contain water chemically bound within their crystalline structure; these are referred to as hydrates.

Glossary xxiii

Hypertension:
Raised blood pressure.

Hypnotic:
A substance that induces sleep.

Impurity Profiling:
The characterisation of naturally occurring or synthetic by-products in a drug to form a "fingerprint" that may be characteristic of its origin or manufacturing route.

INN:
International Nonproprietary Name. Defined by the World Health Organisation (**WHO**) for substances that have or have had therapeutic value.

IUPAC:
International Union of Pure and Applied Chemistry. A body responsible for the systematic nomenclature of chemical entities.

Legalisation:
The complete removal of a conduct or activity from the criminal and civil law.

Leuckart route:
A popular method for converting ketones (*e.g.* **P2P** and **PMK**) to the corresponding **amines** using formic acid, ammonium formate or formamide/methylformamide as reagents. When the ketone is P2P, the result is amphetamine or methylamphetamine, while MDA and MDMA arise from the ketone PMK.

Marquis test:
A field test using a reagent consisting of 10% formaldehyde in concentrated sulfuric acid. Various colours are produced by mixing the reagent with different drugs.

Mass spectrum:
A pattern of charged molecular fragments produced by bombarding molecules with electrons. The fragmentation pattern is characteristic of the molecule.

Narcolepsy:
A disease causing the patient to fall asleep at unpredictable times.

Narcotic analgesic:
A type of **analgesic** acting on the central nervous system rather than on peripheral nerves. Many **opioids** (*e.g.* diamorphine) are typical narcotic analgesics.

Neurotransmitter:
A chemical messenger involved in passing a signal from one neuron to adjacent neurons in the brain. These include serotonin, dopamine, glutamate and γ-aminobutyric acid (GABA).

Nitrogenous base:
A synthetic or naturally occurring substance containing one or more amine nitrogen atoms in its structure and acting as a **base**.

Noradrenaline:
Also known as norepinephrine. An example of a **neurotransmitter**, it is a naturally occurring substituted **phenethylamine**. Substances that interact with the noradrenaline receptor are said to be noradrenergic.

Opiate:
One of a group of alkaloids isolated or chemically derived from opium. Often synonymous with **opioid**.

Opioid:
One of a group of alkaloids isolated or chemically derived from opium. Often synonymous with **opiate**, but sometimes restricted to semi-synthetic products (*e.g.* diamorphine) or chemically related synthetic substances (*e.g.* methadone).

Opium:
The dried latex of the seed capsule of the opium poppy (*Papaver somniferum* L.).

P2P:
1-phenyl-2-propanone: a ketone often used as a precursor to amphetamine and methylamphetamine. Also known as phenylacetone and benzylmethylketone (BMK).

Phenethylamine:
Phenethylamine is 2-phenylethylamine. The term is used less precisely to mean a derivative of phenethylamine, often by substitution in the side-chain or in the aromatic ring or both. The phenethylamine family includes a range of substances that may be **stimulants**, **entactogens** or **hallucinogens**.

PMK:
Piperonylmethylketone, a ketone often used as a precursor in the manufacture of MDMA. Also known as 3,4-methylenedioxyphenyl-2-propanone.

Glossary

Potency:
A quantitative measure of the activity or strength of a drug: a different concept to **purity**, which is the proportion of active drug in a preparation.

Primary amine:
A chemical group comprising a nitrogen atom attached to two hydrogen atoms and to a carbon atom.

Psychoactive drug:
A substance that affects the mind or mental processes. The term is often used in a broad sense to include both **psychotropic** and **narcotic** drugs.

Pychotomimetic:
A drug, such as LSD, that mimics the effects of psychosis such as sensory hallucinations. Now less commonly used than **hallucinogen**.

Psychotropic drug:
A generic term for substances that modify normal behaviour. It includes *inter alia*, stimulants, hallucinogens, tranquillisers, hypnotics. To a certain extent, the term has become synonymous with those substances listed in the Schedules of the UN 1971 Convention.

Purity:
The proportion (%) of active drug in a preparation: a different concept from **potency**. Most laboratories determine purities with respect to the **base** because in a sample sent for analysis, the particular salt form cannot be determined without further, often unnecessary, investigation. So, for example, pure amphetamine base has a purity defined as 100%. When amphetamine base reacts with, *e.g.*, sulfuric acid to form the sulfate salt, then the purity of that salt, with respect to the base, is 79%; the remaining 21% is the sulfate residue. If the purity is expressed with respect to a specific salt form, then pure amphetamine sulfate has a purity of 100%.

***R*-enantiomer:** See enantiomer.

Racemate, also Racemic mixture:
A 50:50 mixture of two **enantiomers** produced when a synthesis is not stereoselective.

Reclassification:
The process of moving a substance from one Class to another in the (UK) Misuse of Drugs Act, and the legislation of certain other countries.

Reduction:
A chemical process involving removal of oxygen atoms and/or addition of hydrogen atoms.

Reductive amination:
A chemical process involving removal of oxygen atoms and addition of amino groups.

S-enantiomer: See enantiomer

Salt:
The product of reacting a base with an acid. Many salts are soluble in water but insoluble in organic solvents.

Secondary Amine:
A chemical group comprising a nitrogen atom attached to a hydrogen atom and two carbon atoms.

Serotonin:
Also known as 5-hydroxytryptamine. An example of a **neurotransmitter**, it is a naturally occurring substance closely related to synthetic hallucinogenic **tryptamines**.

Simon test:
A field test using a reagent consisting of sodium carbonate, acetaldehyde and sodium nitroprusside. Used for distinguishing **primary amines** from **secondary amines**.

Stereoisomer:
One of two or more forms of a molecule with the same sequence of atoms, which arises from the three-dimensional arrangement of those atoms.

Stimulant:
A substance that increases psychomotor activity, often by increasing the production of certain **neurotransmitters** in brain synapses.

Tachycardia:
Raised pulse rate.

Tertiary amine:
A chemical group comprising a nitrogen atom attached to two or three carbon atoms but bearing no hydrogen atoms.

THC:
Δ^9-Tetrahydrocannabinol, the major active principle in cannabis.

Tryptamine:
Tryptamine itself is 1H-indole-3-ethanamine, but the term is also used less precisely to mean a derivative of tryptamine, often by substitution at the side-chain nitrogen atom or in the aromatic ring or both. The tryptamine family includes numerous **hallucinogens** and/or substances that interact with **serotonin** receptors.

CHAPTER 1
Introduction

1.1 DRUG MISUSE

Drugs whose possession or supply is restricted by law are known as scheduled or, in the UK, as controlled substances. They comprise both licit materials (*i.e.* those manufactured under licence for clinical use) and the illicit products of clandestine factories. Although many plant-based drugs have been self-administered for thousands of years (*e.g.* coca leaf, cannabis, opium, and peyote cactus), the imposition of criminal sanctions is mostly a product of the 20th century. Many of the drugs currently abused were once not only on open sale, but often promoted as beneficial substances by the food and pharmaceutical industries. A pattern developed whereby initial misuse of pharmaceutical products such as heroin, cocaine and amphetamine led to increasing legal restrictions and the consequent rise of an illicit industry. Nowadays, most serious drug abuse involves illicit products. Most fall into just a few pharmacological groups, *e.g.* central nervous system stimulants, narcotic analgesics, hallucinogens and hypnotics. The most prevalent of these are the plant-derived or semi-synthetic substances (*e.g.* cannabis, cocaine and heroin), but the view of the former United Nations Drug Control Programme is that wholly synthetic drugs (*e.g.* amphetamine, MDMA and related designer drugs) are likely to pose a more significant social problem in the future. There is an increasing recognition of the problems caused by misuse of medicinal products, primarily benzodiazepine tranquillisers. In a risk-assessment process carried out in the UK by the Advisory Council on the Misuse of Drugs (ACMD; see Chapter 11), benzodiazepines were rated as

Forensic Chemistry of Substance Misuse: A Guide to Drug Control
By L.A. King
© L.A. King 2009
Published by the Royal Society of Chemistry, www.rsc.org

Table 1.1 Drug use in the past year, 16–59 year-olds, England and Wales[3].

Drug	% Population	Class in Misuse of Drugs Act
Cannabis	8.2	Class B pending (Class C in survey period)
Cocaine powder	2.6	Class A
Ecstasy	1.8	Class A
Amyl nitrite	1.4	Not controlled
"Amphetamines"	1.3	Class B (mostly amphetamine)
"Magic mushrooms"	0.6	Class A
"Tranquillisers"	0.4	Class C (Benzodiazepines)
Ketamine	0.3	Class C
Crack cocaine	0.2	Class A
LSD	0.2	Class A
Glues	0.2	Not controlled
Heroin	0.1	Class A
Anabolic steroids	0.1	Class C
Any drug	10.0	n/a

more harmful than any of the other Class B or Class C drugs examined except the barbiturates. Mortality from drug abuse is largely associated with opiates[1]. Thus, in 2004 in the United Kingdom, heroin or morphine was mentioned on 971 death certificates, methadone on 280, cocaine on 185 and ecstasy on 66.

On the basis of a recent Home Office report (Drug Misuse Declared: Findings from the 2006/07 British Crime Survey – see Bibliography), a third of the adult population in the United Kingdom (UK) admits to having used a controlled drug at least once in their lives; fewer than 10% use drugs on a regular basis and for the great majority of these the drug involved is cannabis. Table 1.1 shows the proportion of 16–59 year-olds who admit to using a specific drug in the past year. After cannabis, the next most commonly used drugs are cocaine and 3,4-methylenedioxymethylamphetamine (MDMA; ecstasy). Seizure data from police and customs show a broadly similar pattern. In 2007/8, there were over 228 000 drug offences[2] in the UK, the majority of which involved cannabis. In Europe, it is estimated that 0.2–0.3% of the population are regular heroin users. With few exceptions, the scale of drug abuse has steadily increased in most countries, but is still predominantly associated with younger members of the population.

---

[1] *United Kingdom Drug Situation, 2007 Edition, UK Focal Point on Drugs, Annual Report to the European Monitoring Centre for Drugs and Drug Addiction* – see Bibliography

[2] C. Kershaw, S. Nicholas and A. Walker, Crime in England and Wales 2007/08: *Findings from the British Crime Survey and police recorded crime*, Home Office, 2008: http://www.homeoffice.gov.uk/rds/pdfs08/hosb0708.pdf

[3] *Drug Misuse Declared, 2006/7* – see Bibliography

Drugs seized by law-enforcement agencies and suspected to be controlled (scheduled) substances are normally submitted to a forensic science laboratory for formal identification, and, where appropriate, quantification. In certain circumstances (the guilty plea policy), this rule is relaxed provided that a number of conditions are met and the substances (amphetamine, cocaine, heroin and morphine) have been provisionally identified by a field test (Appendix 11).

Profiles of the major drugs of misuse (amphetamine, cannabis, cocaine and crack cocaine, heroin, LSD, MDMA and methylamphetamine) are provided in Appendix 10. Typical purities and common adulterants are listed in Appendix 12, while typical street prices and wrap (*i.e.* street deal) sizes are shown in Appendix 13.

1.2 ABBREVIATIONS

In the abbreviations and acronyms listed below, only the more frequently mentioned drug substances are included. Acronyms for many phenethylamines and tryptamines can be found in the publications PIHKAL and TIHKAL respectively, (see Bibliography). In this book, the Misuse of Drugs Act 1971 is referred to as the "Act". Although "MDA" is used in some publications as an abbreviation for Misuse of Drugs Act, this is not ideal since MDA is also the acronym for 3,4-methylenedioxyamphetamine – one of the ecstasy drugs. The Misuse of Drugs Regulations, 2001 are shown as the "Regulations". Substances listed in Schedule 2 to the Act are correctly known as "controlled drugs", but, for the sake of clarity and when the context is clear, are often described herein simply as "drugs" or "substances". By normal convention, the term "mg" is used in the following text for milligrams(s). It may be noted that, until the 2001 revision, the Regulations used the obsolete term milligrammes. Some well-known abbreviations, such as THC and MDMA will not be found in the Misuse of Drugs Act because they are subsumed by generic definitions. Thus, THC is *"a tetrahydo derivative of cannabinol"*, and MDMA is *"a compound . . . structurally derived from . . . an N-alkylphenethylamine . . . by substitution in the ring . . . with . . . alkylenedioxy . . . substituents"*. LSD is listed specifically under the approved name lysergide. Although cannabis and cannabis resin are distinct entities, it is sometimes convenient to describe them both under the collective term "cannabis". Similarly, "cannabinols" is used to mean cannabinol and cannabinol derivatives. A distinction is only made in the text where legal or chemical aspects need to be identified. According to the World Health Organisation (WHO), scheduled drugs are "abused" rather than "misused", but in the following text the two terms are used synonymously. The Misuse of Drugs Act and other UK legislation uses the word "sulphate", whereas the preferred term in the international literature is now sulfate. This book uses sulfate, sulfuric, *etc.*, except where the legislation is directly quoted. Expanded definitions of some abbreviations shown below can be found in the Glossary.

1,4-BD:	1,4-Butanediol
2C-I:	2,5-Dimethoxy-4-iodophenethylamine
2C-T-2:	2,5-Dimethoxy-4-ethylthioamphetamine
2C-T-7:	2,5-Dimethoxy-4-propylthiophenethylamine
4-MTA:	4-Methylthioamphetamine
ACMD:	Advisory Council on the Misuse of Drugs
API:	Active Pharmaceutical Ingredient
BAN:	British Approved Name
BZP:	1-Benzylpiperazine
CAS:	Chemical Abstracts System
CBD:	Cannabidiol
CBN:	Cannabinol
CND:	Commission on Narcotic Drugs (a UN body)
CNS:	Central Nervous System
DPMA:	Drugs (Prevention of Misuse) Act 1964
ECDD:	Expert Committee on Drug Dependence (part of WHO)
EMCDDA:	European Monitoring Centre for Drugs and Drug Addiction
EMEA:	European Medicines Evaluation Agency
EU:	European Union
EWS:	Early Warning System (EMCDDA)
GBL:	γ-Butyrolactone
GHB:	γ-Hydroxybutyrate
INN:	International Nonproprietary Name
IUPAC:	International Union of Pure and Applied Chemistry
LSD:	Lysergide; Lysergic acid diethylamide
MBDB:	*N*-Methyl-1-(1,3-benzodioxol-5-yl)-2-butanamine
MDA:	3,4-Methylenedioxyamphetamine
MDEA:	3,4-Methylenedioxyethylamphetamine
MDMA:	3,4-Methylenedioxymethylamphetamine
MDPEA:	Methylenedioxyphenethylamine
MHRA:	Medicines and Healthcare Products Regulatory Agency
NPAS:	New Psychoactive Substance (EMCDDA) post-2005
NSD:	New Synthetic Drug (EMCDDA) pre-2005
P2P:	1-Phenyl-2-propanone

PEA:	Phenethylamine
PIHKAL:	Book: Phenethylamines I Have Known and Loved
PMK:	3,4-Methylenedioxyphenyl-2-propanone
PMA:	4-Methoxyamphetamine
SI:	Statutory Instrument
THC:	Δ^9-Tetrahydrocannabinol
THCA:	Δ^9-Tetrahydrocannabinolic acids
TIHKAL:	Book: Tryptamines I Have Known and Loved
TMA-2:	2,4,5-Trimethoxyamphetamine
UN1961:	United Nations Single Convention on Narcotic Drugs (1961)
UN1971:	United Nations Convention on Psychotropic Substances (1971)
UN1988:	United Nations Convention Against Illicit Traffic in Narcotic Drugs and Psychotropic Substances (1988)
WHO:	World Health Organisation

CHAPTER 2
Control of Chemical Substances

2.1 INTRODUCTION

Much has been written about the pros and cons of whether drugs of misuse should be controlled by the State, decriminalised or even legalised. It is not the purpose of this book to rehearse those arguments, but rather to take a broader view and accept that society does need to control some chemicals to protect individuals and society at large from their harmful effects. Controlled drugs are then part of a continuum, which is not necessarily one-dimensional and where no clear demarcation lines can be drawn. Many of those who advocate legalisation of controlled drugs would still see a need for controls elsewhere. The difference between individual attitudes is then mostly a question of where they sit in that continuum. The purpose of this chapter is to examine, mostly from a European perspective, the legal controls on a wide range of other chemical substances. Later, in Chapter 12, the question is raised as to whether these various and unrelated controls can be consolidated into a unified scheme, which at the same time might rationalise some of the existing problems with the Misuse of Drugs Act.

In the following sections, the classification is not exhaustive. Thus, attention is given to poisons, medicines, drug precursors, chemical weapons and their precursors, solvents and gases, and the "social" drugs. But many other legal controls exist. Thus, there are restrictions by HM Revenue and Customs on the importation of alcohol, tobacco and many other substances. These do not arise because of their chemical properties or harmfulness, but are rather a reflection of the fact that

---

Forensic Chemistry of Substance Misuse: A Guide to Drug Control
By L.A. King
© L.A. King 2009
Published by the Royal Society of Chemistry, www.rsc.org

various duties may be payable (*e.g.* excise duty). However, importation restrictions are sometimes imposed in order to protect domestic industry. An example here is saccharin, which at various times and places has been considered harmful in an economic sense to sugar production[1]. In sporting events, controls exist on a wide range of stimulants, glucocorticosteroids, alcohol, anabolic steroids, diuretics and other substances[2], but these are beyond the scope of the current analysis.

2.2 POISONS

To a certain extent, the concept of a poison underlies some of the modern controls on drugs. In the 16th century, Paracelsus famously noted that *"All substances are poisons; there is none which is not a poison. The right dose differentiates a poison"*. In modern usage, a poison is defined as a chemical that interferes with living functions by permanently blocking an essential biological process, with the capability of causing death.

It is instructive to note that in the 1960s, the UK attitude to drugs of misuse was still partly framed by experience with the regulation of poisons. The increasing misuse of amphetamine, related stimulants and hallucinogens in the early 1960s could not be controlled by the Dangerous Drugs Acts. An attempt was made in 1964 to prosecute under the Pharmacy and Poisons Act 1933 in a case (R-v-Esam and Page) involving supply of lysergide (LSD). However, that Act, which pre-dated the discovery of LSD, only referred to *"Ergot, alkaloids of; their homologues"*. Following dissent on what was meant by terms such as homologue and alkaloid, whether lysergide is a homologue of lysergamide, whether lysergamide occurs naturally in ergot and whether ergot subsumes *Claviceps paspali* as well as *Claviceps purpurea*, the case was dismissed. The Government then referred the problem to the Poisons Board, a statutory body set up by that Act, which had been designed to control the sale or supply to the public of medicines and a wide range of potentially dangerous substances. This led to The Drugs (Prevention of Misuse) Act, 1964, but the link with poisons had not been broken; the title to this Act and later Modification Orders would still refer to "Poisons". In Australia, a strong link between poisons and controlled

---
[1] C.M. Merki, *Sugar Versus Saccharin: Sweetener Policy before World War I*, in *The Origins and Development of Food Policies in Europe*, ed. J. Burnett and J.O. Derek, Leicester University Press, 1994, pp 192–202
[2] The World Doping Agency (WADA) maintains a list of substances that are prohibited by athletes and other sportsmen and women before or during sporting events at national and international level: http://www.wada-ama.org/en/prohibitedlist.ch2

substances is evidenced by the title of the appropriate legislation: Drugs, Poisons and Controlled Substances Act 1981.

The current UK legislation is the Poisons Act 1972 and the Poisons Rules 1982 (as amended). The Act was designed to guard against the misuse by accident, inadvertence or criminal design of nonmedicinal poisons to which the public need to have access. Substances considered as poisons are listed in the Poisons Rules. According to Section 2 of the Act, Part 1 poisons may only be sold by a pharmacist, whereas Part II poisons may be sold by a seller who has registered with the local authority. The Act sets out the requirement for storage, packaging, labelling and the documentation of all sales. Some of the listed poisons can only be sold in certain types of preparations for agricultural and related uses. The substances[3] covered by Part I and Part II of the Poisons Rules include *inter alia*, certain "organophosphorus" compounds and other pesticides, salts of arsenic, barium and mercury, mineral acids, nicotine, paraquat and formic acid.

2.3 PHARMACEUTICAL INGREDIENTS AND MEDICINAL PRODUCTS

To a large extent, and provided they do not fall into one of the other categories discussed in this chapter, there are few restrictions on the manufacture, sale or possession of an active pharmaceutical ingredient (API). For most purposes, each is treated as if it were just another chemical; the fact that APIs have a pharmacological effect on a user is incidental. However, once an API is converted into a medicinal product then it becomes subject to the Medicines Act, 1968.

This was introduced following a review of legislation prompted by the thalidomide tragedy in the 1960s. It brought together most of the previous legislation on medicines and introduced a number of other legal provisions for the control of medicines. The Act divided medicinal drugs into three categories, depending mainly on the dangers they posed and the risk of misuse. The categories are:

- Prescription Only Medicines (POM), which may be sold or supplied to the public only on a practitioner's prescription. They may be administered only by or in accordance with directions from an appropriate practitioner; a term that includes a medical practitioner.

---

[3] An example of the full list of poisons and other requirements can be found at: http://www.luton.gov.uk/internet/business/business_and_street_trading_licences/non_medical_licences/Licence%20-%20non%20medicinal%20poisons

With the exception of controlled drug preparations below a certain strength set out in Schedule 5 to the Misuse of Drugs Regulations 2001, all controlled drugs with medical use are Prescription Only Medicines.
- Pharmacy medicines (P), which, subject to certain exceptions, may be sold or supplied only from registered premises by, or under the supervision of, a pharmacist. Most products listed in Schedule 5 of the Misuse of Drugs Regulations 2001 are pharmacy medicines.
- General sales list medicines (GSL), which may be sold or supplied direct to the public in an unopened manufacturer's pack from any lockable premises. No controlled drugs are general sales list medicines.

The original definition of a medicinal product was set out in Section 130 (1) of the Medicines Act 1968 as: " . . . *any substance or article (not being an instrument, apparatus or appliance) which is manufactured, sold, supplied, imported or exported for use wholly or mainly in either or both of the following ways, that is to say*:

(a) *use by being administered to one or more human beings or animals for a medicinal purpose;*
(b) *use, in circumstances to which this paragraph applies, as an ingredient in the preparation of a substance or article which is to be administered to one or more human beings or animals for a medicinal purpose.*"

As a general rule, a medicinal product means a recognisable dosage unit, for example a tablet, capsule, skin patch, injection ampoule or sublingual spray. But a tablet, *etc.*, only becomes a medicinal product if it contains " . . . *any substance or combination of substances which may be used in or administered to human beings either with a view to restoring, correcting or modifying physiological functions by exerting a pharmacological, immunological or metabolic action* . . . "[4]. A grey area exists where some manufacturers of illicit psychoactive drugs may sell them on the basis that they are not for human consumption. This occurs particularly with some of the piperazine derivatives and other "new" drugs (Chapter 9), which may be described by the suppliers as, for example, "plant growth stimulators". Other problems that can arise with the definition of a medicinal product are discussed in Chapter 8.

---
[4] Article 1.2 of Directive 2004/27/EC

2.4 DRUG PRECURSORS

With the exception of those drugs that are used in their naturally occurring state (*e.g.* cannabis) or are diverted from legitimate clinical sources, most illicit drugs of abuse require the use of chemicals either to facilitate their extraction from natural products, or to form semi- or fully synthetic substances. The Convention Against Illicit Traffic in Narcotic Drugs and Psychotropic Substances (UN1988), as later modified, now includes 23 such chemicals. Trade controls on these have been introduced by most member States and have been extended by some. Within the European Union, Regulation (EC) No. 273/2004 establishes harmonised measures for intra-Community control and monitoring. It contains provisions relating to licences, customer declarations and labelling. The corresponding rules for monitoring trade in such chemicals between the Community and third countries are set out in Regulation (EEC) No. 111/2005. In UK domestic law, certain activities with precursor chemicals are controlled by the Criminal Justice (International Cooperation Act 1990) as modified, the Controlled Drugs (Drug Precursors) (Intra-Community Trade) Regulations 2008 (S.I. 2008/295) and the Controlled Drugs (Drug Precursors) (Community External Trade) Regulations 2008 (S.I. 2008/296). This legislation includes, for example, a list of countries where individual export authorisation is required from the Home Office, as well as export registration requirements for certain annual threshold amounts of Category 3 chemicals (see below).

Apart from reagents such as mineral acids and solvents that are used on a large scale primarily in cocaine processing, the essential precursor chemicals listed include those often used to manufacture amphetamine (*i.e.* 1-phenyl-2-propanone, phenylacetic acid), methylamphetamine (ephedrine and pseudoephedrine), lysergide (ergotamine, ergometrine, lysergic acid), MDMA and related drugs (safrole, isosafrole, piperonal, 3,4-methylenedioxyphenyl-2-propanone), heroin (acetic anhydride) and methaqualone (anthranilic acid, *N*-acetylanthranilic acid). These and other chemicals may be recovered from suspect shipments or from the scenes of illicit drug synthesis and submitted for laboratory analysis.

Appendix 5 lists the precursors and other essential reagents that are set out in UK legislation. Category 1 chemicals are those regarded as true precursors, that is to say they form the core structure of the product drug. Category 2 chemicals are considered to be secondary precursors; they are either convertible into Category 1 precursors or are used as essential reagents. The materials in Category 3 are mostly acids and solvents, used as adjuncts in drug processing. In general terms, the

2.5 CHEMICAL WEAPONS AND THEIR PRECURSORS

The Chemical Weapons Act 1996 sets out the restrictions on the use, possession and manufacture of certain agents that may be used as weapons or are precursor chemicals to those agents[5]. The need for this legislation derived from the United Nations Chemical Weapons Convention, which came into force in 1997. The proscribed substances fall into two groups: (A) toxic chemicals, which include variously substituted phosphonofluoridates, phosphoramidocyanidates and phosphonothiolates, sulfur mustards, nitrogen mustards and lewisites, and (B) precursor chemicals to those toxic chemicals.

legitimate industrial uses and consumption of these 23 chemicals are least for Category 1 and greatest for Category 3.

2.6 SOLVENTS AND GASES

Nonaqueous liquids are widely used as solvents in synthetic processes or as cleaning agents. Those used in drug manufacture, and under international control, are described in Appendix 5. Ethanol is discussed later, while γ-butyrolactone (GBL) and 1,4-butanediol (1,4-BD) are covered in Chapter 9. Because of their volatility and physiological effects, certain solvents present special risks when inhaled; these are described below.

2.6.1 Volatile Solvents and Gases

The most commonly abused volatile solvents comprise low molecular weight alkanes (*e.g.* butane from cigarette lighter fuels), toluene (a solvent in some glues) and various aerosol propellants, all of which are readily obtained from domestic products. Even though these chemicals continue to be associated with fatal poisonings, particularly in young people, they are ubiquitous and it is unlikely that they could ever be brought within the scope of the Act. However, some controls on their sale do exist. For example, the Intoxicating Substances (Supply) Act 1985 makes it an offence for a retailer to sell solvents to anyone under the age of 18, knowing that they are being purchased to be abused. It does not make it illegal to buy or own solvents. The Cigarette Lighter

---

[5] http://www.opsi.gov.uk/acts/acts1996/ukpga_19960006_en_1

Refill (Safety) Regulations 1999 – an amendment to the Consumer Protection Act 1987 – makes it illegal to supply gas cigarette lighter refills to anyone under the age of 18. Furthermore, European Directive 2005/59/EC of 26th October 2005 prohibits the placing on the market, for sale to the general public, the substance toluene and adhesives and spray paints containing in excess of 0.1% toluene. Member States should have applied these measures from 15th June 2007.

Alkyl nitrites form a distinct subgroup of volatile solvents. Although amyl nitrite has recognised value as a coronary vasodilator and antidote to cyanide poisoning, illicit products (so-called "poppers") used to contain isobutyl nitrite. However, The Dangerous Substances and Preparations (Safety) Regulations 2006 prohibit the sale of isobutyl nitrite, largely because it has been shown to be a human carcinogen[6]. Manufacturers of poppers now use isopropyl nitrite, which is believed to have similar physiological effects to its homologues. Alkyl nitrites are unlikely subjects for control under the Act. Their status under the Medicines Act, particularly if they contain nitrites other than amyl nitrite, has been a contentious issue. An attempted prosecution under the Medicines Act in 1999 against a supplier (Quietlynn Ltd) of poppers containing isobutyl nitrite was unsuccessful because the defence was able to show that these products did not cause significant harm.

Apart from low molecular weight alkanes, some true gases are abused. In moderate amounts, nitrous oxide (laughing gas) causes intoxication. It was once used as an anaesthetic agent, but because of the high partial pressures needed has long since been superseded. Even modern anaesthetic gases are not without their risks, and it may be noted that halothane (shown as Галотан) is listed as a scheduled substance in List 3 (Psychotropic Substances) of the Russian drug code[7]. Helium has been used to facilitate suicides[8].

2.7 THE "SOCIAL" DRUGS

2.7.1 Alcohol

Along with caffeine and tobacco, alcohol is one of the most widely used psychoactive substances; its use or prohibition in different countries is

---

[6] http://poppers.cfsites.org/custom.php?pageid = 8068
[7] W.E. Butler, *HIV/Aids and drug misuse in Russia: Harm reduction programmes and the Russian Legal System*, International Family Health, 2003
[8] V. Auwaerter, M. Perdekamp, J. Kempf, U. Schmidt, W. Weinmann and S. Pollak, *Toxicological analysis after asphyxial suicide with helium and a plastic bag*, For. Sci. Int., 2007, **170(2–3)**, 139–141

partly a reflection of cultural attitudes and partly a recognition of commercial interests. The situation is further complicated by evidence that suggests that small amounts of alcohol may be of therapeutic value for some people. In the risk assessments carried out by ACMD (Chapter 11), alcohol (ethanol, C_2H_5OH) was given an overall score of 1.85 out of a possible 3. This placed it above amphetamine and close to "street" methadone. Alcohol is undoubtedly a harmful substance, and its hazards are often cited by those who would wish to see cannabis and other drugs legalised. In the UK it is estimated that less than 10% of the adult population never consume alcohol[9]. But it is often forgotten that alcohol is subject to many legal controls. Using the UK experience as an example, it cannot be sold without a licence[10], or sold to anyone below the age of 18, its consumption in many public places is prohibited by local bye-laws, and it is an offence to distil alcohol. Furthermore, the Road Traffic Act 1988 sets limits on how much alcohol may be present in a driver's blood, breath or urine before an offence is committed.

2.7.2 Tobacco

In 2005, 24 per cent of adults aged 16 or over in Great Britain smoked cigarettes[11]. In the ACMD scale of drug harm, tobacco scored 1.62 out of a possible 3. This placed it well above Class A drugs such as LSD and ecstasy. Tobacco is an example of a plant product that contains a psychoactive drug, viz. nicotine, where the intact vegetable substance is more harmful than the active constituent. Although nicotine [S-3-(1-methylpyrrolidin-2-yl)pyridine; Structure (2.1)] is an extremely addictive and toxic substance, the hazards of tobacco arise largely from the fact that it is smoked; the acute and chronic harms are caused by tars and other substances. In 1996, the World Health Organisation Expert Committee on Drug Dependence (ECDD) considered a review of nicotine, but it has remained unregulated by the UN Conventions. In the meantime, nicotine is available over the counter in the form of gum (*e.g.* Nicorette ®) and other products to assist those wishing to quit smoking. The sale of tobacco is licensed, and it may not be sold to those under the age of 18 years[12]. In recent years, social pressure against smoking has increased. This has included prohibition of many forms of advertising, and a ban on smoking in public buildings, pubs and restaurants.

---

[9] Institute of Alcohol Studies Factsheet, http://www.ias.org.uk
[10] Licensing Act 2003: http://www.opsi.gov.uk/acts/acts2003/ukpga_20030017_en_1
[11] http://www.statistics.gov.uk/cci/nugget.asp?id = 866
[12] The Children and Young Persons (Sale of Tobacco, *etc.*) Order 2007; http://www.opsi.gov.uk/si/si2007/uksi_20070767_en_1

Proposals are in hand to require retailers not to display cigarettes, to ban tobacco vending machines and other measures[13]. Tobacco represents an example of a hazardous substance where methods other than direct legislation have been effective in curbing use.

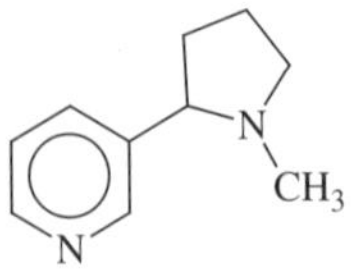

Structure (2.1) Nicotine

2.7.3 Caffeine

More ubiquitous than even alcohol or tobacco, caffeine (3,7-dihydro-1,3,7-trimethyl-(1H)-purine-2,6-dione; Structure (2.2)) is well known as a stimulant mostly derived from tea, coffee or guarana. But, unlike amphetamine and cocaine, which are strongly dopaminergic, caffeine and the closely related theobromine (3,7-dihydro-3,7-dimethyl-(1H)-purine-2,6-dione), a constituent of cocoa, exert their effects by interaction with adenosine receptors, although a side effect is increased levels of dopamine and noradrenaline (norepinephrine).

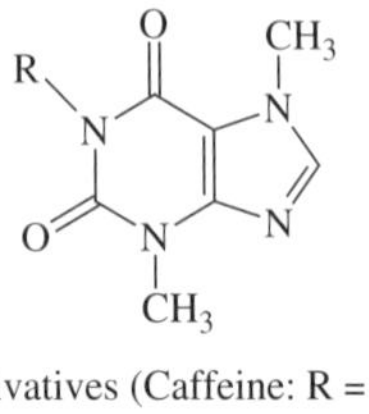

Structure (2.2) Xanthine derivatives (Caffeine: R = CH_3; Theobromine: R = H)

Fatalities have occurred following the ingestion of 5 to 50 g caffeine but these two xanthine derivatives must rank as the least harmful drugs, so much so that, in the popular consciousness, they are probably even less regarded as drugs than either alcohol or tobacco. However, there is some evidence that caffeine can produce addiction in some individuals as shown by the appearance of withdrawal symptoms following cessation of use. Unpublished work carried out for the Independent Enquiry into the Misuse of Drugs Act (see Bibliography) included a quantitative risk

---

[13] http://www.the-tma.org.uk/page.aspx?page_id = 43

assessment of 18 substances by a group of psychiatrists. Caffeine was ranked as the least harmful substance, although its score was not zero. Despite having no nutritional value, coffee and tea are generally regarded as foodstuffs, with no more controls than apply to other foods. There also appear to be few cultural prohibitions in any country on their consumption. However, caffeine tablets would normally be regarded as medicinal products. Caffeine is probably the most common adulterant found in illicit amphetamine powders in Europe. It serves as a cheap diluent that can extend the physiological effects of amphetamine.

2.7.4 Khat

Khat (also known as qat or chat) comprises the leaves and fresh shoots of *Catha edulis*, a flowering evergreen shrub cultivated in East Africa and the Arabian Peninsula. The active components, S-cathinone [(−)-2-aminopropiophenone; Structure (2.3)] and 1S,2S-cathine [(+)-norpseudoephedrine; Structure (2.4)], are usually present at around 0.3 to 2.0%. Both substances are close chemical relatives of synthetic drugs such as amphetamine and methcathinone[14].

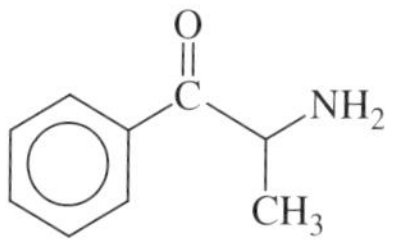

Structure (2.3) Cathinone

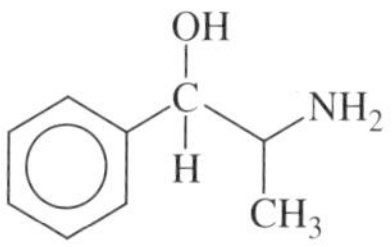

Structure (2.4) Cathine

Khat is scheduled in some European countries, the US and Canada, but not in the UK. Following a recent review of khat by ACMD, it was suggested that some form of licensing of premises where khat is consumed might be applicable, along similar lines to the licensing for sale of alcohol. However, for legal reasons this was considered inappropriate because khat contains controlled substances. Whereas khat may be free of domestic controls, it cannot be legally exported to those countries

---

[14] K. Szendrei, *The chemistry of khat*, Bull. Narcotics, 1980, **32(3)**, 5–35

where it is controlled. Although *Catha edulis* is not under international control, cathine and cathinone are listed in UN1971 under Schedules III and I, respectively; in the UK they are both Class C substances. Alcoholic extracts (tinctures) of khat have been noted especially in "Herbal High" sales outlets and at music festivals. The only known prosecution under the Act for the unlawful production of cathine and cathinone from khat was unsuccessful (R-v-Farmer, Lewes Crown Court, 1998). At the 34th meeting of ECDD in 2006, no recommendation was made for scheduling khat under the international drug control conventions.

2.8 DANGEROUS SUBSTANCES

There is much Health and Safety legislation that is concerned in a general way with chemical hazards, including explosives[15], but that is beyond the scope of the present book. However, The Dangerous Substances and Preparations (Safety) Regulations 2006 – part of consumer protection legislation – are specifically concerned with the sale of certain dangerous materials. For example, with a number of stated exceptions, there is a prohibition on the sale of: toluene, adhesive or spray paint containing more than 0.1% toluene; substances and preparations containing more than 0.1% benzene or certain chlorinated solvents; certain carcinogens, mutagens and substances with reproductive toxicity[16]; a child's dressing gown that has been treated with tris(2,3-dibromopropyl) phosphate, tri(aziridin-1-yl)phosphine oxide or polybrominated biphenyls; lachrymatory products or those designed to induce sneezing; toys and childcare items containing more than 0.1% of various phthalates. As discussed earlier, these Regulations also ban the sale of products containing isobutyl nitrite.

Much more comprehensive controls on a huge number of chemicals have been foreshadowed in EU legislation[17]. The "Reach" project puts the onus on business rather than public authorities for safety testing, and will include the thousands of chemicals that have been used for years without proper understanding of their effect on health or the environment.

---

[15] http://www.hse.gov.uk/explosives/information/licencereg.htm
[16] Schedule 2 of The Dangerous Substances and Preparations (Safety) Regulations 2006 lists a large number of substances that fall under the categories of carcinogens, mutagens and substances with reproductive toxicity
[17] http://ec.europa.eu/environment/chemicals/reach/reach_intro.htm

CHAPTER 3
Nomenclature

3.1 BRITISH APPROVED NAMES AND INTERNATIONAL NONPROPRIETARY NAMES

In common with most chemical substances, a given drug may have a number of synonyms. Wherever possible, the Act uses the British Approved Name (BAN), which is defined by the British Pharmacopoeia Commission. In general, a BAN only exists for drugs that have or have had clinical value. If no BAN exists then the formal chemical name may be used, *i.e.* a name in agreement with the rules of the International Union of Pure and Applied Chemistry (IUPAC).

However, European Law (Directives 65/65 and 92/27/EEC) requires the use of an International Nonproprietary Name (INN) for the labelling of medicinal products. In 2003, a decision was made to abandon most BANs and use International Nonproprietary Names in the British Pharmacopoeia. The assignment of an INN to a substance is decided by the WHO. In many cases, the differences between a BAN and an INN are minor. Table 3.1 sets out those controlled drugs where the name used in the Act differs from the INN.

There is no intention at present to ensure that the names of drugs in the Act should be changed to correspond with International Nonproprietary Names. Apart from the administrative burden, several difficulties would arise. For example, there are many occasions where the INN refers to a specific stereoisomer, even though the names in the Act must necessarily include all stereoisomers. Furthermore, at least one hybrid name exists in the Act. Thus, *N*-hydroxy-tenamphetamine is not

Table 3.1 Controlled drugs where the name used in the Act differs from the International Nonproprietary Name (INN)[a]

Name in Act	Class	INN
4-Bromo-2,5-dimethoxy-α-methylphenethylamine	A	Brolamfetamine
Dimenoxadole	A	Dimenoxadol
N-Hydroxy-tenamphetamine	A	N-Hydroxytenamfetamine
Methadyl acetate	A	Acetylmethadol
Tilidate	A	Tilidine
Amphetamine	B	Amfetamine
Methylamphetamine	B	Metamfetamine
Methylphenobarbitone	B	Methylphenobarbital
Quinalbarbitone	B	Secobarbital
Benzphetamine	C	Benzfetamine
Clorazepic acid	C	Dipotassium clorazepate
Diethylpropion	C	Amfepramone
N-Ethylamphetamine	C	Etilamfetamine
Ethyloestrenol	C	Ethylestrenol
Fenethylline	C	Fenetylline
Methandienone	C	Metandienone
Methenolone	C	Metenolone
Methyprylone	C	Methyprylon
Stanolone	C	Androstanolone

[a] Amfepramone and tilidine are proposed International Nonproprietary Names. N-Hydroxytenamfetamine is not an INN, but this construction would seem appropriate by analogy to other "amfetamines". Metamfetamine is strictly the INN for the d-isomer only. Dipotassium clorazepate is a salt of clorazepic acid. Since salts are already subsumed in the Act, it would be more logical to retain the name as clorazepic acid.

a BAN, an INN or an acceptable IUPAC name. Even the core word "Tenamphetamine" itself is an anglicised version of "Tenamfetamine": the International Nonproprietary Name for one of the ecstasy drugs commonly called 3,4-methylenedioxyamphetamine (MDA).

3.2 SYNONYMS AND COMMON TERMS

Table 3.2 gives examples of drug names that either do not occur in the Act as such because of generic definitions or there is a better-known abbreviation or the trivial name is more widely used. In other cases, US English offers alternative spellings or there are acceptable chemical synonyms. A large number of slang terms for drugs are in use although their popularity varies from place to place and in time; and a few are shown in Table 3.2. Apart from "Bromo-STP" and "STP", a large number of other controlled phenethylamines (mainly the so-called PIHKAL compounds) are invariably described by acronyms (Appendix

Table 3.2 Synonyms and other names for certain controlled drugs.

Common name	Name in Act	Alternative name	Slang terms
Bromo-STP, DOB	4-bromo-2,5-dimethoxy-α-methyl-phenethylamine	Brolamfetamine; 4-bromo-2,5-dimethoxyamphetamine	
Cannabis	Cannabis	Marijuana (US), hemp	} Pot, Dope, Blow, Weed
Cannabis resin	Cannabis resin	Hashish (US)	} Ganga, Charas
Heroin[a]	Diamorphine	Diacetylmorphine	"H", Horse, Skag, Smack
N-Hydroxy MDA	N-Hydroxy-tenamphetamine	N-Hydroxytenamfetamine	
LSD	Lysergide	Lysergic acid diethylamide	LSD-25, Acid
MDEA	(Generically subsumed)	3,4-methylenedioxyethylamphetamine	"E", Eve, Ecstasy
MDMA	(Generically subsumed)	3,4-methylenedioxymethylamphetamine	"E", Adam, Ecstasy, XTC
Methcathinone	Methcathinone	Ephedrone	Crank
Methylamphetamine	Methylamphetamine	Methamphetamine (US), Metamfetamine (INN for the d-isomer)	Meth, Speed, Ice
Pethidine	Pethidine	Meperidine (US)	
STP	2,5-dimethoxy-α,4-dimethyl-phenethylamine	2,5-dimethoxy-4-methylamphetamine	

[a]Heroin and diamorphine are not strictly synonymous. The former is a crude preparation obtained by the acetylation of morphine, in which diamorphine is often the major component.

17). A further example of synonymy occurs in the generic definitions where halide [paragraph 1(c)] and halogeno [paragraph 1(d) of Part I of Schedule 2] should be interpreted as having the same meaning.

3.3 REDUNDANCY

The introduction, from the late 1970s, of generic definitions based on chemical substitution patterns (Chapter 6) led to a certain amount of duplication. In other words, some controlled drugs continued to be named specifically, but also fell within the scope of generic control. However, this situation was not new; several examples of redundancy can be traced back to the 1972 Protocol, which revised UN1961. Amongst others, this extended controls to include the esters or ethers of substances in Schedule I. The best example of added redundancy is that both morphine and heroin (diamorphine) continued to be listed by name. However, the explicit retention of diamorphine, the diacetyl ester of morphine, was not strictly required. It can be argued that no harm is caused by this duplication. Indeed removal of the word "Diamorphine", while having no legal significance, might be misunderstood and lead to unnecessary debate. Apart from the diamorphine/morphine pair, a number of other opioids continued to be listed in Schedule I of UN1961 and as Class A controlled drugs in the Act even though they were either ethers or esters of other listed substances. A further level of duplication is caused by the retention in Schedule I of UN1961, and in Part I of Schedule 2 to the Act, of three racemic forms. Thus, racemoramide, racemethorphan and racemorphan could be deleted because they each contain 50% of an existing controlled drug, namely dextromoramide, levomethorphan and levorphanol, respectively. Table 3.3 shows examples of Class A substances where effective double entry occurs in the Act.

3.4 THE MEANING OF "DERIVATIVE" IN THE MISUSE OF DRUGS ACT

The concept of a derivative enters the legislation in several places; and the examples of cannabinol, ecgonine, lysergamide and pentavalent nitrogen morphine derivatives are described in Chapter 6. Since all organic compounds could be described as derivatives of methane, one of the simplest of all carbon-based compounds, then it follows that they are all derivatives of each other. This *reductio ad absurdum* shows that the word must be given a much tighter meaning when interpreting the Act. It is now usually accepted amongst forensic chemists that compound **A** is a

Nomenclature

Table 3.3 "Double entry" of Class A controlled drugs in the Act.

Specific name (Paragraph 1(a) of Part I of Schedule 2)	Generic control in Part I of Schedule 2
Bufotenine	}
N,*N*-Diethyltryptamine	} Paragraph 1(b) – as ring-substituted tryptamines
N,*N*-Dimethyltryptamine	}
Psilocin	}
4-Bromo-2,5-dimethoxy-α-methylphenethylamine	}
2,5-Dimethoxy-α-4-dimethyl phenethylamine	} Paragraph 1(c) – as ring-substituted phenethylamines
Mescaline	}
Alfentanil	}
Carfentanil	} Paragraph 1(d) – as fentanyl derivatives
Lofentanil	}
Sufentanil	}
Allylprodine	}
Alphameprodine	}
Alphaprodine	} Paragraph 1(e) – as pethidine derivatives
Properidine	}
Trimeperidine	}
Diamorphine	Paragraph 3 – as a di-ester of morphine
Hydrocodone	Paragraph 3 – as an ether of hydromorphone
Levomethorphan	Paragraph 3 – as an ether of levorphanol
Methadyl acetate	Paragraph 3 – as an ester of methadone (enol form)
Myrophine	Paragraph 3 – as a di-ester of morphine
Nicomorphine	Paragraph 3 – as a di-ester of morphine
Oxycodone	Paragraph 3 – as an ether of oxymorphone
Thebacon[a]	Paragraph 3 – as an ester of hydrocodone (enol form)

[a]Thebacon is an ester and an ether of hydromorphone and is therefore not covered directly by paragraph 3 of Part I of Schedule 2. In other words, hydrocodone and thebacon are not both redundant, but either could be listed without the other

derivative of compound **B** only if **B** can be converted to **A** *in a single chemical reaction,* even if that is only achievable in a theoretical sense.

3.5 DIALKYL DERIVATIVES

In Modification Order S.I. 3932 of 2001, two phenethylamine derivatives are included (Structures (A19.18) and (A19.19) in Appendix 19) where the nitrogen atom is disubstituted with alkyl groups. This was necessary because some doubt resided in whether *N*,*N*-disubstitution on the amine is currently subsumed by the generic definition in paragraph

1(c) of Part I of Schedule 2 (viz. "...*an N-alkylphenethylamine*..."). However, another instance of "*N*-alkyl" substitution arises in paragraph 1(a) of Schedule 2 Part I. But here the explicit use of the phrase "*Lysergide and other N-alkyl derivatives of lysergamide*", by focusing on lysergide (a dialkyl derivative), was therefore intended to mean that "*N*-alkyl" subsumes *N,N*-dialkyl.

3.6 THE MEANING OF "STRUCTURALLY DERIVED FROM"

A different concept of "derivative" occurs in some of the group-generic definitions discussed in Chapter 6. For example, in ring-substituted phenethylamines, reference is made to *"a compound . . . . . . structurally derived from phenethylamine"*. In the judgement in the case of R-v-Couzens and Frankel in 1992 (Appendix 8), it was accepted that to say that compound **A** is *structurally derived* from **B** does not necessarily mean that **B** can be chemically converted to **A** in one or even several reaction stages. What is meant in the example of phenethylamines is that **A** still contains the carbon skeleton of phenethylamine (*i.e.* **B**), but that additional atoms (carbon, oxygen or other as defined) are now attached without implying that such an attachment is chemically possible. In practical terms, it will almost always be the case that **A** and **B** are produced from quite separate precursor chemicals that, in this example, may not in themselves be phenethylamines.

3.7 HOMOLOGUES

A particular type of derivative is known as a homologue. This term is used to describe a member of a series of chemical substances where each member differs from the next by a constant structural feature. The simplest example occurs with the straight-chain alkanes. Thus, methane, ethane, n-propane, n-butane *etc.* are part of a homologous series where the constant difference between adjacent members is a methylene (CH_2) moiety. It is therefore correct to say, for example, that ethane, n-propane, n-butane, *etc.* are the higher homologues of methane and that methane and ethane are the lower homologues of n-propane. The term homologue can be found in the legislation in respect of cannabinol and its tetrahydro derivatives. Thus, in Part IV of Schedule 2, these controlled derivatives are described as "*3-alkyl homologues*". Since the Act does not define "homologues", there is some contention on whether this phrase means both higher and lower homologues. The more common view is that only the higher homologues are included, in which case the cannabivarins with a 3-propyl group are not controlled.

3.8 "PHENETHYLAMINES", "PHENYLETHYLAMINES", PHENYLALKYLAMINES AND "AMPHETAMINES"

The generic definition embedded in paragraph 1(c) of Part I of Schedule 2 to the Act refers to certain substitution patterns in phenethylamine. The term "phenethylamine" is a contraction for 2-phenylethylamine (also known as β-phenylethylamine; Structure (3.1)); it does not refer to the isomeric 1-phenylethylamine (also known as α-phenylethylamine; Structure (3.2)). Because of this distinction, derivatives of 1-phenylethylamine (Chapter 9), even when they otherwise satisfy the substitution pattern set out in paragraph 1(c) of Part I of Schedule 2, are therefore not controlled drugs.

Structure (3.1) 2-Phenylethylamine   Structure (3.2) 1-Phenylethylamine

Many phenethylamines of interest are α-methyl-substituted and it is common practice to refer to them in a non-IUPAC approved shorthand form as amphetamine (viz. α-methylphenethylamine) derivatives. Similarly, N,α-dimethyl-substituted phenethylamines are often named as methylamphetamine derivatives. Phenylalkylamine is a more general term for substances where a phenyl group is attached to a carbon atom in an alkylamino group, where the alkyl moiety contains any number of carbon atoms. The Misuse of Drugs Act uses the terms amphetamine and methylamphetamine. In many countries, methylamphetamine is known as methamphetamine. In UN1971, the terms amfetamine and metamfetamine are used.

CHAPTER 4
Drug Control at International and European Level

In international law, controls on drugs of abuse are set out in three United Nations (UN) Treaties: The Single Convention on Narcotic Drugs 1961 (UN1961), the Convention on Psychotropic Substances 1971 (UN1971) and The Convention Against Illicit Traffic in Narcotic Drugs and Psychotropic Substances 1988 (UN1988). These Treaties are implemented in domestic laws by signatories, and have been considerably extended in some States. In the UK, the corresponding legislation is the Misuse of Drugs Act 1971 and the Misuse of Drugs Regulations 2001, as amended. Since the inception of the UN Conventions, numerous substances have been added to the Schedules, particularly those of the 1971 Treaty.

4.1 UNITED NATIONS SINGLE CONVENTION ON NARCOTIC DRUGS (1961)

In the 1961 Convention[1], there is a strong emphasis on plant-based drugs, with rules for their cultivation, manufacture and trade. The drugs are set out in four Schedules, the principal objectives of which are to provide proportionate levels of restriction on legitimate trade in the drugs. Those restrictions follow the order I > II > III. The substances in Schedule I (by far the largest group) include cannabis, opium and

---

[1] *Single Convention on Narcotic Drugs 1961*, International Narcotics Control Board: http://www.incb.org/incb/convention_1961.html

---

Forensic Chemistry of Substance Misuse: A Guide to Drug Control
By L.A. King
© L.A. King 2009
Published by the Royal Society of Chemistry, www.rsc.org

cocaine[2]. In addition, there are around 100 synthetic narcotic analgesics, but only a few of these are now used clinically or ever reported to be abused. Schedule II includes, for example, codeine and dihydrocodeine, Schedule III includes preparations of certain narcotic drugs with rules on the maximum concentrations or amounts that may be present. An example here is codeine, which falls into Schedule III if it is *"compounded with one or more other ingredients and containing not more than 100 milligrams of the drug per dosage unit and with a concentration of not more than 2.5 per cent in undivided preparations"*. Schedule IV is somewhat paradoxical in that it is not less restrictive than Schedule III. It includes a small group of substances (*e.g.* cannabis, cannabis resin and heroin) that are already listed in Schedule I, but that are considered to be particularly dangerous, and to which special provisions should apply. All signatories to the Convention have incorporated the listed substances and general principles of control into their domestic law, but most have chosen to realign and expand the Schedules such that those controls are listed in a more logical order. For example, in the UK Misuse of Drugs Regulations 2001 and the US Controlled Substances Act there are five Schedules where controls decrease in the order I to V.

4.2 UNITED NATIONS CONVENTION ON PSYCHOTROPIC SUBSTANCES (1971)

More than 100 psychotropic substances are listed in the 1971 Convention[3], but again only a small fraction is regularly abused. The term "psychotropic" is not defined in the Convention. There are four Schedules of controlled substances, ranging from Schedule I (most restrictive) to Schedule IV (least restrictive). Schedule I includes drugs that are believed to be most dangerous and whose therapeutic value is doubtful, *e.g.* LSD, N,N-dimethyltryptamine and THC. In Schedule II are drugs such as amphetamine, where some limited therapeutic value is recognised. Schedule III includes, for example, cathine and buprenorphine, while Schedule IV includes pemoline, aminorex and benzodiazepines. Thus, Schedule I largely comprises illicit substances, whereas Schedules II, III and IV include legitimate pharmaceutical agents (APIs). A notable feature is that barbiturates are spread across Schedules II, III and IV. Part of the

---

[2] In modern usage, the word narcotic is usually confined to the naturally occurring and synthetic opioids such as morphine and methadone and related compounds. Both cannabis and cocaine would now be described as psychotropic drugs.
[3] *Convention on Psychotropic Substances 1971*, International Narcotics Control Board: http://www.incb.org/incb/convention_1971.html

reason is that secobarbital, the only example in Schedule II, was identified as having a particularly high fatal toxicity[4], whereas phenobarbital (Schedule IV) not only has a much lower toxicity and is not a hypnotic, but is used in the treatment of epilepsy. The inclusion of tetrahydrocannabinol (THC) and its isomers in Schedule I has led to some confusion since cannabis, cannabis resin and extracts of cannabis are listed in the 1961 Convention.

Unlike the Treaty of 1961, there was originally no overarching control of the stereoisomers of psychotropic drugs (see Chapter 6 for a discussion of stereoisomerism). Thus, in Schedule I, amfetamine (meaning both the "–" and the "+" enantiomers) is listed together with dexamfetamine (the "+" enantiomer of amphetamine) and levamfetamine (the "–" enantiomer) while metamfetamine (meaning the "+" enantiomer) is listed alongside metamfetamine racemate (a mixture of the "–" and "+" enantiomers). These examples, and the situation whereby the stereochemical configuration of many other substances was left unspecified, led to some confusion. The UN has since clarified the status of stereochemical variants in the 1971 Convention[5]. These problems were avoided in the Misuse of Drugs Act by the inclusion of the stereoisomers of almost all controlled drugs (Chapter 6).

4.3 UNITED NATIONS CONVENTION AGAINST ILLICIT TRAFFIC IN NARCOTIC DRUGS AND PSYCHOTROPIC SUBSTANCES (1988)

The purpose of the 1988 Convention was to provide additional legal mechanisms for enforcing the 1961 and 1971 Conventions. The treaty is concerned with tackling organised crime, by co-operation in tracing and seizing drug-related assets. To limit money laundering, it allows signatories to empower their courts to seize bank and commercial records. Whereas there had hitherto been no clear requirement for signatories to criminalise drug possession, this Convention is more explicit on the need for this. According to the text of Article 3 of the 1988 Convention: *"Subject to its constitutional principles and the basic concepts of its legal system, each Party shall adopt such measures as may be necessary to establish as a criminal offence under its domestic law, when committed intentionally, the possession, purchase or cultivation of narcotic drugs or*

---

[4] L.A. King and A.C. Moffat, *Hypnotics and sedatives: An index of fatal toxicity* Lancet, 1981, **(i)**, 387–388

[5] *Green List, Annex to the annual statistical report on psychotropic substances (Form P), 23rd edition*, August 2003, INCB, Vienna

psychotropic substances for personal consumption contrary to the provisions of the 1961 Convention, the 1961 Convention as amended or the 1971 Convention". There has been some debate about the precise interpretation of Article 3.

Article 12 established two groups of precursor chemicals, with the objective of limiting their diversion into illicit drug manufacture (Appendix 5).

4.4 EUROPEAN INITIATIVES

Sitting between national legislations and the international drug control treaties, the European Union (EU) has a supranational role. Apart from precursor legislation, which derives from the United Nations Convention Against Illicit Traffic in Narcotic Drugs and Psychotropic Substances 1988, the EU has specific competence in the area of "new psychoactive substances", formerly known as "new synthetic drugs".

4.4.1 The Period 1997 to 2005

From the early 1990s, many so-called "designer drugs" were regularly discovered in the European Union. They were often psychotropic substances related to amphetamine and MDMA. Their appearance raised questions about possible health risks and the problems that could arise if such substances were arbitrarily controlled in some Member States, but not in others. It was agreed that progress could be made by sharing information and by establishing a risk-assessment procedure and a mechanism for their eventual EU-wide control. This led to the "Joint action concerning the information exchange, risk assessment and control of new synthetic drugs" (NSD)[6]. These substances were defined as those that had a limited therapeutic value and were not at that time listed in the 1971 UN Convention on Psychotropic Substances, yet posed as serious a threat to public health as the substances listed in Schedules I and II to that Convention. Thus, well-established drugs such as amphetamine, MDMA and its ethyl homologue (MDEA) were excluded since they were already controlled in international law. The term "new" did not refer to newly invented, but rather "newly misused"; most of the drugs in question were first created many years ago. The

---

[6] Adopted by the Council of the European Union under the Dutch Presidency on 16 June 1997 (Official Journal L 167, 25 June 1997)

"Joint action" took place against a political background whereby Europe had become a leading producer of synthetic drugs.

Under the 1997 "Joint action", over 30 NSD were reported through the Early Warning System (EWS). Some were identified in biological samples, but most were found in police or customs seizures. However, few occurred in large amounts or were widespread, and most had a limited life on the illicit market. The drugs identified since 1997 were largely phenethylamines (mostly listed in the book "PIHKAL"), tryptamines (mostly listed in "TIHKAL") and, less commonly, substituted cathinones and substituted piperazines. Risk assessments[7] were carried out on nine of them (MBDB, 4-MTA, GHB, ketamine, PMMA, TMA-2, 2C-T-2, 2C-T-7 and 2C-I). Although neither ketamine nor GHB (γ-hydroxybutyrate) strictly qualified as "new synthetic drugs", it was considered appropriate to carry out risk assessments because at that time there was information of misuse, but they were not under international control. A common feature of the remaining seven drugs was that they were often found as tablets marked with "illicit" logos similar to those seen on ecstasy tablets, but that provided no clue to their chemical contents. By contrast, the notified tryptamines, none of which has so far been risk assessed, were more commonly seen as powders. Of the above nine substances, 4-MTA, PMMA, TMA-2, 2C-T-2, 2C-T-7 and 2C-I were brought under control throughout the EU. Subsequently, 4-MTA and GHB were added to Schedules I and IV, respectively, of the 1971 UN Convention on Psychotropic Substances (Table 4.1).

4.4.2 Developments since 2005

In 2002, a review of the Joint action was carried out under the terms of the EU Action Plan on Drugs 2000–2004. Suggestions for improvements led to a process that, in 2005, culminated in Council Decision 2005/387/JHA[8]. This broadened the scope of, and replaced, the 1997 "Joint action", while maintaining a three-step approach. The term NSD was replaced with "new psychoactive substance" (NPAS), which included not only substances that might qualify for inclusion in UN1971, but also narcotic substances that would normally be listed in UN1961. Furthermore, there

---

[7] EMCDDA, *Report on the risk assessment of MBDB in the framework of the joint action on new synthetic drugs, June 1999, 4-MTA*, October 1999, GHB, June 2002, *ketamine*, June 2002, *PMMA*, March 2003, TMA-2, May 2004, (*2C-I, 2C-T-2 and 2C-T-7*), May 2004 http://www.emcdda.europa.eu/index.cfm?fuseaction = public.Content&nNodeID = 431&sLanguageISO = EN

[8] The Council Decision was published in the Official Journal (L 127/32-37) on 20 May 2005

Table 4.1 Substances risk assessed by EMCDDA and their classification[a] under the Misuse of Drugs Act and UN1971.

Substance	Risk-assessment report by EMCDDA	Proposed for control in EU	Class in Misuse of Drugs Act	Schedule in UN1971
MBDB	1999	No	A	Not listed
4-MTA	1999	Yes	A	I
GHB	2002	No	C	IV
Ketamine	2002	No	C	Not listed
PMMA	2003	Yes	A	Not listed
TMA-2	2004	Yes	A	Not listed
2C-I	2004	Yes	A	Not listed
2C-T-2	2004	Yes	A	Not listed
2C-T-7	2004	Yes	A	Not listed
BZP	2007	Yes	C	Not listed

[a]Apart from 4-MTA, all other Class A substances fall within the generic definition of a substituted phenethylamine (Chapter 6). GHB was subsequently added to UN1971. Ketamine was later controlled under the Misuse of Drugs Act (Class C). BZP was the first substance to be recommended for control within the European Union that was not already controlled by the Misuse of Drugs Act.

was no restriction to synthetic drugs, so that naturally occurring products would also qualify for investigation.

In the information exchange/Early Warning stage, once a NPAS is detected on the European market, Member States ensure that information on the manufacture, traffic and use of the drug is transmitted to the EMCDDA and Europol. If a risk assessment is carried out then the Council may decide that the drug should be controlled. Appropriate measures and criminal penalties in the EU Member States are decided in line with national laws, which in turn comply with the UN Conventions. The Council decision does not prevent individual Member States from unilaterally introducing national control measures at an earlier stage if they consider it appropriate.

CHAPTER 5
Drug Legislation in the UK

5.1 HISTORICAL BACKGROUND

Apart from the Pharmacy Act of 1868, which restricted the sale of opium, the modern period of drug control started in the early 20th century. Following the Poisons and Pharmacy Act 1908 and the Shanghai Opium Commission in 1909, further restrictions were introduced on cocaine, morphine and opium. More controls on a wider range of substances were introduced by the successive Dangerous Drugs Acts of 1920, 1925 and 1951. By 1951, there was a simple list of "traditionally abused drugs" that included opium, opiates, cocaine and cannabis. The UN Single Convention of 1961 required that this short list be expanded to cover a large group of substances. These were mostly synthetic opiates and other narcotic analgesics, but many were, and remain to this day, as little more than chemical curiosities. Inclusion of the UN1961 substances in UK legislation led to the Dangerous Drugs Act of 1964. An opportunity was then taken to redefine cannabis as no longer just the flowering tops of female plants, and, for the first time, to create an offence of cannabis cultivation. These changes were consolidated in the Dangerous Drugs Act of 1965.

In the meantime, new problems appeared that had not been anticipated by UN1961. Stimulants such as amphetamine and related compounds and other psychotropic drugs rapidly became more prevalent, and special legislation was needed for their control. This became the Drugs (Prevention of Misuse) Act (DPMA) 1964. Following legal difficulties that arose in a 1964 prosecution under the Pharmacy and Poisons Act of

Forensic Chemistry of Substance Misuse: A Guide to Drug Control
By L.A. King
© L.A. King 2009
Published by the Royal Society of Chemistry, www.rsc.org

1933 (see Chapter 2), which concerned supply of LSD, a Modification Order to the DPMA was made in 1966. This introduced specific control to mescaline and lysergamide and extended control to N-alkyl derivatives of lysergamide and ring-hydroxy derivatives of N,N-dimethyltryptamine. Problems arising from the meaning of "derivative" first arose at this time. This issue, which was discussed in a 1973 publication[1], is further expanded in Chapter 3.

The first attempts to introduce structure-specific generic control into UK drugs law were made with the DPMA of 1964. This contained a statement intended to cover a range of synthetic stimulants. The precise text (in italics) read: *"Any synthetic compound structurally derived from either α-methylphenethylamine [i.e. amphetamine] or β-methylphenethylamine by substitution in the side chain, or by ring closure therein, or by both such substitution and such closure, except . . . "* [named substances]. While this did indeed encompass compounds such as phentermine and methylphenidate, it soon became clear that a refined interpretation of the generic statement unwittingly included dozens of drugs that were not stimulants[2]. In fact, it could be argued that some barbiturates such as phenobarbitone were also covered. Difficulties arose with the interpretation when multiple bonds were present in the side chain or substitution by oxidation occurred in the side chain. The generic control was repealed by a Modification Order in 1970. Following this early failure, it would be some years before generic control of phenethylamines again entered the legislation. But this time (1977), the focus would be on ring-substituted phenethylamines, it would be much more robust and would be followed by generic controls for several other groups.

5.2 THE MISUSE OF DRUGS ACT 1971

The Misuse of Drugs Act 1971, which replaced the Dangerous Drugs Act 1965 and the Drugs (Prevention of Misuse) Act 1964, introduced the concept of "controlled drugs". These are defined as those substances or products set out in Parts I, II and III of Schedule 2, and are shown here in Appendix 15. The meaning of certain terms is set out in Part IV of Schedule 2 (Appendix 16). The Act, which came into effect in 1973, set up an Advisory Council on the Misuse of Drugs whose terms of reference included a definition of what should constitute a controlled drug. This is set out in Section 1(2) of the Act as: *"It shall be the duty of the Advisory*

---

[1] G.F. Phillips, *When is a derivative not a derivative?*, Med. Sci. Law, 1973, **13(3)**, 216–220
[2] G.F. Phillips, *Controlling drugs of abuse*, Chemistry in Britain, 1972, 123–130

Council to keep under review the situation in the United Kingdom with respect to drugs which are being or appear to them likely to be misused and of which the misuse is having or appears to them capable of having harmful effects sufficient to constitute a social problem . . . ". The ACMD succeeded an earlier body: The Advisory Committee on Drug Dependence.

The Misuse of Drugs Act coincided with, and incorporated the changes introduced by, the UN1971 Convention on Psychotropic Substances. The Act prohibits certain activities with respect to controlled drugs (*e.g.* possession, possession with intent to supply and production) without a licence. With the exception of opium, there is no illegality in using or consuming a controlled drug. The great majority of arrests for offences under the Act involve possession of relatively small amounts of a controlled drug. The drugs are listed in Schedule 2 of the Act and are divided into three groups: Class A (Part I of Schedule 2), Class B (Part II) and Class C (Part III). In principle, these groups represent, in decreasing order A to C, the propensity of the substances to cause social harm. Associated with each type of prohibition are the maximum penalties for offences involving controlled drugs, again decreasing in the order A to C. For Class A drugs, the maximum penalty for some offences is life imprisonment, for Class B it is 14 years. For Class C drugs, the maximum penalty for supplying had been five years, but this was increased to 14 years imprisonment by the Criminal Justice Act 2003 (Appendix 7). Penalties for importation/supplying large amounts of cannabis and other drugs are described in Appendix 9.

The list of substances in Schedule 2 may be varied by a Statutory Instrument (S.I.) known as a Modification or Amendment Order. There have been 18 such Orders since 1971 (Appendix 1), most of which have served to incorporate changes agreed by member states of the UN.

Many of the substances listed in the Act and Regulations, numerous definitions and some parts of the generic controls derive directly from the UN Conventions. However, the Act goes beyond the minimum in several important areas. Not only are there more substances, but an important feature of the Act is the extensive use of structure-specific generic terms. Table 5.1 lists drugs by name or group that are controlled in the UK, but are not listed in UN1961 or UN1971.

The structure-specific definitions discussed in Chapter 6 are found only in the Act (and the Regulations) and the Misuse of Drugs Act of the Republic of Ireland 1977 (as amended). A slightly different set of structure-specific definitions of phenethylamines and certain other chemical groups occur in the drugs legislation of New Zealand (Chapter 10). Appendix 7 provides a summary of other legislation concerned with drug control.

Drug Legislation in the UK

Table 5.1 Substances or groups controlled in the UK, but not listed in UN1961 or UN1971.

Substance	Class (Misuse of Drugs Act)
Bufotenine	A
Cannabinol	A
Carfentanil	A
Lofentanil	A
Lysergamide	A
34 Phenethylamines	A
α-Methylphenylhydroxylamine	B
Anabolic steroids and related substances	C
1-Benzylpiperazine and piperazine derivatives	C (Pending)
Chlorphentermine	C
Ketamine	C
Mephentermine	C
Other generically defined groups	A and B

5.2.1 The Misuse of Drugs Act – Substances Removed or Reinstated

In the past thirty years, the scope of the Act has increased to encompass many new substances. Indeed, the generic definitions discussed in Chapter 6 theoretically cover an infinite number. Yet in all this time, only two drugs, namely propylhexedrine and prolintane, have been permanently removed from control. Dexamphetamine (Class B) was also removed as a named substance in 1985 (S.I. 1995), but only because by then it fell to control as a stereoisomer of amphetamine. Substances that have been reclassified since 1971 are described in Chapter 11.

Of the hundreds of named substances in the Act, most are never abused. One reason for their retention is that they are still listed in UN1961 or UN1971 and few substances have ever been removed from international control. A second reason lies closer to home and is well illustrated by the history of pemoline. This anorectic drug was controlled by the Misuse of Drugs Act (and was originally included in the DPMA of 1964) because of its potential for abuse, but licensed products containing pemoline were no longer available by the early 1970s in the UK. Pemoline was therefore removed from the Act in 1973. Within a few years, illicit manufacturers began to produce pemoline tablets. Eventually, pemoline was brought under the scope of Schedule IV of UN1971 and pemoline was reintroduced into the Act in 1989. Apart from pemoline, phentermine and fencamfamin (both of which had also been listed in the DPMA of 1964) were removed only to be later reinstated (Table 5.2). The moral of this episode is that it is probably safer to leave substances under control even when they cease to be an immediate social problem. Unless the pharmaceutical industry should wish to reactivate "old" drugs, their retention causes no problems.

Table 5.2 Substances that have been removed or reinstated since 1971.

Substance	Original control	Removed	Reinstated
Fencamfamin	1971 Act Class C	1973 (S.I. 771)	1986 (S.I.2230) Class C
Pemoline	1971 Act Class C	1973 (S.I. 771)	1989 (S.I. 1340) Class C
Phentermine	1971 Act Class C	1973 (S.I. 771)	1985 (S.I. 1995) Class C
Prolintane	1971 Act	1973 (S.I. 771)	N/A
Propylhexedrine	1986 (S.I. 2230)	1995 (S.I. 1966)	N/A

Because of commitments to the UN, there is in any case only limited scope to remove further substances. Two obvious candidates stand out in the Act, namely chlorphentermine and mephentermine. They are both in Class C and were originally brought under UK control in the DPMA of 1964. Neither is listed in UN1971 and both have long ceased to be available in medicinal products in the UK. However, as with pemoline, their removal could send a signal to clandestine manufacturers.

5.2.2 The "PIHKAL" List

The generic controls on phenethylamines (Chapter 6), which were introduced in 1977, were remarkably far-sighted and comprehensive. Not only did they successfully anticipate the major ecstasy drugs such as MDMA and its congeners, but the generic definition subsumed nearly all of the many ring-substituted amphetamine-type compounds that would appear in UK drug seizures over the next 25 years. The publication of a book in 1991 known as PIHKAL (see Bibliography) and its subsequent appearance on the world-wide web alerted clandestine chemists to the possibilities of creating further designer drugs based on the phenethylamine nucleus. That book provides synthetic methods for over 170 substances with notes on their effects and doses. Of these, over thirty were not covered by the generic definition of 1977. Since many of the "PIHKAL" drugs were broadly similar to the well-known ecstasy compounds and related hallucinogens, the Government recognised that they presented a potential social problem. Although consideration had been given to extending the generic definition of phenethylamines, two arguments worked against this approach. Firstly, the noncontrolled substances formed a heterogeneous set of chemical structures, the inclusion of which would have needed an elaborate definition. This then risked including substances of current or future interest to the pharmaceutical industry. Secondly, the generic control of 1977 has worked remarkably well with no forensic difficulty, but its enlargement could pose a danger not just of incomprehension, but of creating loopholes by virtue of its complexity. In 2001, 34 PIHKAL substances were added to the Act by Statutory Instrument 3932, where they are listed by name as Class A controlled drugs. Although some of these substances were probably inactive or had other properties that would have made them undesirable to users, it was considered a safe precaution to list them all. However, the list does not include the parent compound, phenethylamine itself, nor, through an apparent oversight, 2-methoxy-4,5-methylenethiooxy-amphetamine (PIHKAL #167). Another ring-substituted phenethylamine (4-methylthioamphetamine: 4-MTA) that had not been listed in PIHKAL was included in S.I. 3932 (2001). More information on 4-MTA can be found in Chapter 5 and Appendices 17–19.

Appendix 17 lists the 34 "PIHKAL" compounds as well as 4-MTA firstly as they appear in the Act under their IUPAC names, then as their less formal names and acronyms. Where appropriate, the IUPAC name has been based on the more common root "3,4-methylenedioxyphenyl-Z", where Z is some substituent or other part of the molecule. However, the rules also permit the use of the alternative root "1-(1,3-benzodioxol-5-yl)-Z". The 35 compounds fall into five structural groups (Appendix 18). Groups 1–3 are divided into those (a) where, according to PIHKAL, positive psychoactive effects may be expected, and those (b), where no effect was detected, the effect was unpleasant or the dose was unacceptably high. The molecular structures of the 35 compounds are set out in Structures (A19.1) to (A19.35) in Appendix 19. For comparison, the substitution pattern required by the generic rules of 1977 is shown in Structure (6.40) (Chapter 6). Although the term "acronym" is used for convenience both here and elsewhere, many of the code names for compounds described in PIHKAL bear a rather obscure relationship to the molecular structure.

5.2.3 Substances Added to the UN Conventions (1961, 1971) and/or the Misuse of Drugs Act since 2002

The status of substances subject to international control and/or to the Misuse of Drugs Act as of April 2002 was described in a previous publication[3]. Table 5.3 summarises those substances added to the UN Conventions and/or the Act in the past six years; further information is set out in the following sections. All substances in Table 5.3 can also be found in the appropriate table in Appendix 15.

---

[3] L.A. King, *The Misuse of Drugs Act: A Guide for Forensic Scientists* – see Bibliography

Table 5.3 Substances added to the UN Conventions (1961, 1971) and/or the Misuse of Drugs Act since 2002[a].

Substance or group of substances	UN Convention and Schedule	Class Misuse of Drugs Act	Modification Order	Schedule in Regulations
Amineptine	UN1971 (IV)	Not yet listed	n/a	n/a
4-Androstene-3,17-dione	Not listed	C	(S.I. 1243)2003	4 (Part II)
5-Androstene-3,17-diol	Not listed	C	(S.I. 1243)2003	4 (Part II)
2C-B	UN1971 (II)	A	(S.I. 1243)1977 (Controlled generically)	1
Dihydroetorphine	UN1961 (I)	A	(S.I. 1243)2003	2
γ-Hydroxybutyrate (GHB)	UN1971 (IV)	C	(S.I. 1243)2003	4 (Part I)
Ketamine	Not listed	C	(S.I. 3178)2005	4 (Part I)
4-MTA	UN1971 (I)	A	(S.I. 3932)2001	1
19-Nor-4-androstene-3,17-dione[b]	Not listed	C	(S.I. 1243)2003	4 (Part II)
19-Nor-5-androstene-3,17-diol[c]	Not listed	C	(S.I. 1243)2003	4 (Part II)
Remifentanil	UN1961 (I)	A	(S.I. 1243)2003	2
Zolpidem	UN1971 (IV)	C	(S.I. 1243)2003	4 (Part I)

[a] Substances added to the Misuse of Drugs Act before 2002 are shown in *The Misuse of Drugs Act: A Guide for Forensic Scientists* – see Bibliography
[b] 19-Nor-4-androstene-3,17-dione is a metabolite of 4-androstene-3,17-dione
[c] 19-Nor-5-androstene-3,17-diol is a metabolite of 5-androstene-3,17-diol

5.2.3.1 Amineptine. Amineptine (7-[(10,11-dihydro-5*H*-dibenzo[*a,d*]-cyclohepten-5-yl)amino]heptanoic acid; Structure (5.1)) is an atypical tricyclic anti-depressant with CNS stimulating effects. It has limited therapeutic usefulness because of its hepatotoxicity, and is rarely used in Europe. But in other countries there have been more reports of its abuse and dependence than for other anorectic stimulants currently in Schedule IV of UN1971. In 2005, amineptine was added to Schedule II of the UN1971 Convention, but has yet to be added to the Misuse of Drugs Act.

Structure (5.1) Amineptine

5.2.3.2 Anabolic Steroids (Derivatives of Androstene). In 1996, forty-eight anabolic/androgenic steroids together with some generic definitions were added to the Act. The substances were essentially those proscribed by the International Olympic Committee (IOC). Four further steroids were subsequently added to the IOC list and to the Act as named Class C substances in 2003 since they fell outside the generic definition. They were 4-androstene-3,17-dione, 5-androstene-3,17-diol, 19-nor-4-androstene-3,17-dione and 19-nor-5-androstene-3,17-diol.

5.2.3.3 4-Bromo-2,5-dimethoxyphenethylamine (2C-B). This hallucinogenic drug (Structure (5.2)), which is described in "PIHKAL", falls within the generic definition of a substituted phenethylamine in the Misuse of Drugs Act. It has been added to Schedule II of UN1971.

Structure (5.2) 4-Bromo-2,5-dimethoxyphenethylamine (2C-B)

5.2.3.4 Dihydroetorphine. Dihydroetorphine (7,8-dihydro-7-α-[1-(*R*)-hydroxy-1-methylbutyl]-6,14-*endo*-ethanotetrahydrooripavine; Structure (5.3)) is a potent opioid receptor agonist, chemically related to the Class A drug etorphine. It is used in some countries for pain relief, but is uncommon in Europe. It is listed in Schedule I of UN1961 and became a Class A controlled drug under the Misuse of Drugs Act in 2003.

Structure (5.3) Dihydroetorphine

5.2.3.5 γ-Hydroxybutyrate (GHB). Gamma-hydroxybutyrate (4-hydroxy-n-butyric acid; γ-hydroxybutyrate; GHB) acts as a central nervous system depressant, and is chemically related to the brain neurotransmitter gamma-aminobutyric acid (GABA). Synonyms of GHB include oxybate, gamma-OH, somatomax, "GBH" and "liquid ecstasy". GHB has been used clinically in some European countries as an anaesthetic drug. Recently, it became a licensed medicine in the UK through a decision made by the European Medicines Evaluation Agency (EMEA) to allow the proprietary preparation Xyrem® to be used in the treatment of cataplexy (a disorder associated with narcolepsy often characterised by excessive daytime sleepiness). GHB is easily manufactured by adding aqueous sodium hydroxide to gamma-butyrolactone (GBL) to leave a weakly alkaline solution.

In September 2000, a risk assessment on GHB was undertaken by the EMCDDA (Chapter 4), but a decision was made not to recommend EU-wide control. However, in early 2001, the UN decided that it should be added to Schedule IV of UN1971. Subsequently, in 2003, GHB became a Class C drug under the Misuse of Drugs Act, subject to the controls of Schedule 4 Part I of the Regulations. Because the term gamma-hydroxybutyrate includes both salts and the free acid, it was considered appropriate to list GHB in the Act as 4-hydroxy-n-butyric acid since its salts would be automatically subsumed. A technical difficulty facing control of GHB is that the conversion of GBL into GHB is reversible as shown in Structure (5.4).

Structure (5.4) Interconversion of gamma-hydroxybutyrate (GHB) and its precursor γ-butyrolactone (GBL). Substructure (1) is γ-butyrolactone (GBL); (2) is γ-hydroxybutyric acid (GHB); (3) is the sodium salt of γ-hydroxybutyric acid (GHB)

The precursor (γ-butyrolactone (GBL), sometimes known by the synonym dihydrofuranone) can be simply recovered from a GHB solution by adding acid to neutralise the sodium hydroxide. It is only necessary to add a small amount of water to GBL to produce a detectable level of GHB. This interconversion occurs naturally in the body and drug users have realised that it is possible to ingest the precursor directly to produce the desired effect. Current plans to control GBL are discussed in Chapter 9.

5.2.3.6 Ketamine. Ketamine (2-(2-chlorophenyl)-2-methylamino-cyclohexan-1-one; Structure (5.5)) is a dissociative anaesthetic drug used in animal and occasional emergency human surgery. It is not strictly hallucinogenic, but causes catalepsis (muscle rigidity) and leaves users feeling detached from their immediate environment. In the UK, ketamine is available for hospital use as injection solutions, but there are no preparations licensed for oral use.

Abuse of ketamine was recognised almost thirty years ago in the US, but it did not come to notice until around 1990 in the UK. At that time, seizures often comprised ampoules of the proprietary preparations Ketalar® and Vetalar® or loose powders that had probably been produced by evaporation of these injection liquids. During the mid- to late 1990s, nearly all illicit ketamine was found in the form of well-made tablets, visually similar to, and often sold as, ecstasy tablets. In the past few years, loose powders have become more common. This may be a reflection of the illicit market responding to the several successful prosecutions of ketamine tablet manufacturers for attempted supply of a controlled drug, *i.e.* a MDMA "look-alike".

Ketamine is closely related to phencyclidine (Structure (5.6)), a Class A controlled drug, and to the veterinary anaesthetic tiletamine (Structure (5.7)). Although tiletamine is not controlled in the UK or internationally, it is listed (as a combination product with zolazepam) in Schedule III of the US Controlled Substances Act. In 2000, a risk assessment was carried out on ketamine by the European Monitoring Centre for Drugs and Drug Addiction (EMCDDA; Chapter 4), but there was insufficient evidence to recommend that it should be controlled by Member States of the European Union.

Following a review by ACMD (Chapter 11), ketamine was recommended for control, and became a Class C drug subject to the provisions of Schedule 4 (Part 1) of the Regulations in 2005. A critical review of ketamine was also carried out at the 34th meeting of ECDD (WHO) in 2006, but no recommendation was made for scheduling under UN1971.

Structure (5.5) Ketamine

Structure (5.6) Phencyclidine

Structure (5.7) Tiletamine

5.2.3.7 α-Methyl-4-(methylthio)-phenethylamine (4-MTA). Although 4-MTA (Structure (A19.35)) was not described in "PIHKAL", it had been reported in the pharmacological literature as a possible anti-depressant drug and had appeared as a new designer drug in Europe. A number of seizures of 4-MTA were made by police and customs across Europe in the following two years. At least five fatal poisonings were recorded in the UK alone where 4-MTA had been the direct or indirect cause of death. This substance was one of several so-called "new synthetic drugs" that were notified under the European Union Joint action (Chapter 4) and subjected to risk assessment. As a consequence of this evaluation, the European Council of Ministers decided in 1999 that 4-MTA should be controlled in all Member States of the Union. In early 2001, the UN CND meeting in Vienna agreed that 4-MTA should be brought within the scope of UN1971 as a Schedule I substance. In the UK, 4-MTA [shown as α-Methyl-4-(methylthio)-phenethylamine] is listed as a Class A controlled drug. Like the "PIHKAL" drugs, it is not covered by the 1977 generic definition, but is named specifically in paragraph 1(ba) of Part I of Schedule 2 of the Act.

5.2.3.8 Remifentanil. Remifentanil {methyl [3-[4-methoxycarbonyl-4-(N-phenylpropanamido) piperidino]propionate]; Structure (5.8)} is an analgesic related to fentanyl. It has a similar potency, but with a short duration of action. It was listed as a Schedule I drug in UN1971 and became a Class A controlled drug under the Misuse of Drugs Act in 2003. It is listed specifically because it fails to meet the fentanyl derivatives generic description (Chapter 6); the phenethyl group has been replaced with a group not specified in the definition.

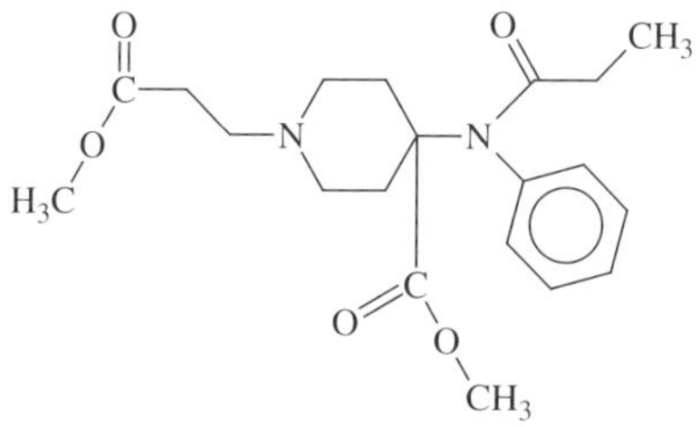

Structure (5.8) Remifentanil

5.2.3.9 Zolpidem. Although zolpidem [*N,N*,6-trimethyl-2-*p*-tolylimidazo[1,2,-α]pyridine-3-acetamide; Structure (5.9)] is chemically unrelated to the benzodiazepines, its pharmacology and abuse potential are broadly similar. It has a short duration of action, and is used as a hypnotic. Following a recommendation by the WHO, it was added to Schedule IV of the 1971 Convention in 2001. It was added to the Misuse of Drugs Act as a Class C drug (Schedule 4, Part I) in 2003. It is available in the UK as the proprietary preparation Stilnoct®.

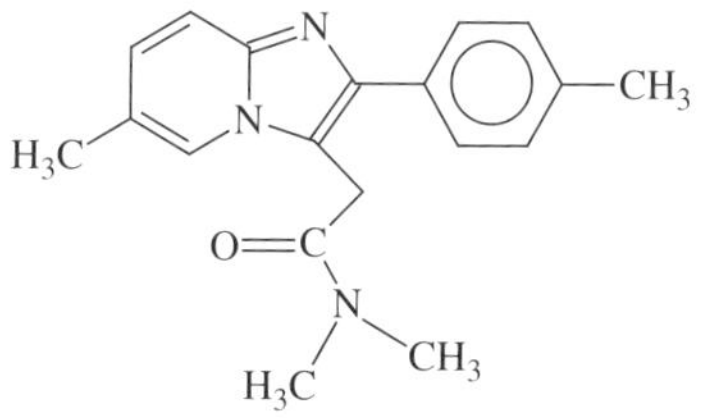

Structure (5.9) Zolpidem

5.3 THE MISUSE OF DRUGS ACT: CHANGES PENDING

As discussed in Chapter 11, cannabis and cannabis resin are to be reinstated in Class B, and remain in Schedule 1 of the Regulations, while 1-benzylpiperazine (BZP), and possibly other generically defined piperazines, will become Class C controlled drugs in Schedule 1 of the Regulations. Amineptine (Schedule IV in UN1971), oripavine (Schedule I of UN1961) and a number of anabolic steroids have yet to be added to the Act (see Chapter 9).

5.4 THE MISUSE OF DRUGS REGULATIONS, 2001

In the Misuse of Drugs Regulations 2001, which came into force on 1st February 2002 and replaced the previous Regulations of 1985, controlled drugs are divided into five Schedules based on a balance between their value as medicines and their hazards as drugs of abuse. In simple terms, the Regulations set out what *should* be done with controlled drugs, whereas the Act sets out what *should not* be done. In broad terms, at least for psychotropic drugs, the Schedules correspond to the respective Schedules of the 1971 Convention. Whereas it is permitted for a substance listed in UN1971 to be placed in a higher Schedule in the Regulations, it may not be placed lower. Table 5.4 shows the relationship for those few substances that appear in a different Schedule in the Regulations, 2001 compared to UN1971. The connection between the Regulations and the Schedules of UN1961 is less precise, although the principle still holds that National Governments must not permit less stringent controls on substances than those set out in UN1961.

Controls are placed on the manufacture, prescription, storage and record keeping of the substances in decreasing order 1 to 5. Drugs in Schedule 1 may not be prescribed, but can be used under licence in medical and scientific research, whereas substances in Schedule 4 Part II, provided they are in the form of a medicinal product, are freely available to the extent that there is no possession offence. Further exceptions to certain offences with some "low-dose" preparations occur in Schedule 5. Table 5.5 gives examples of the two-dimensional matrix of UK drug control. Most Class C drugs are found in Schedule 4 and most Class A drugs are found in Schedules 1 and 2 of the Regulations, but there is otherwise little correlation between the Class of a substance in the Act and its Schedule in the Regulations.

Table 5.4 Substances that appear in a different Schedule in the Misuse of Drugs Regulations 2001 compared to their Schedule in UN1971.

Substance	Schedule (Misuse of Drugs Regulations, 2001)	Schedule (UN1971)
Benzphetamine	3	IV
Diethylpropion	3	IV
Ethchlorvynol	3	IV
Ethinamate	3	IV
Glutethimide	2	III
Lefetamine	2	IV
Mazindol	3	IV
Meprobamate	3	IV
Methylphenobarbitone	3	IV
Methyprylone	3	IV
Phendimetrazine	3	IV
Phentermine	3	IV
Pipradrol	3	IV
Temazepam	3	IV

Table 5.5 Relationship between Class in the Act and Schedule in the Regulations for selected substances.

Regulations	Class A	Class B	Class C
Schedule 1	Lysergide, MDMA	Cannabis (pending), Methcathinone	Cathinone
Schedule 2	Diamorphine, Cocaine	Amphetamine, Codeine	Dextropropoxyphene
Schedule 3	-	Barbiturates (most)	Temazepam, flunitrazepam
Schedule 4 Part I	-	-	Other benzodiazepines[a]
Schedule 4 Part II	-	-	Anabolic steroids
Schedule 5	"Low dose" Morphine	"Low dose" Codeine	"Low dose" Dextropropoxyphene

[a]Midazolam (one of the benzodiazepines) was transferred from Schedule 4 Part I to Schedule 3 in early 2008 by the Misuse of Drugs and Misuse of Drugs (Safe Custody) (Amendment) Regulations 2007 (S.I. 2154). This move was partly prompted by evidence of abuse of midazolam by a medical practitioner

The full text of Schedules 4 and 5 is set out in Appendix 2 and Appendix 3, respectively. A cross-reference to the Schedule in the Regulations and the Class in the Act for all controlled drugs can be found in Appendix 15.

Additions to Schedule 4 introduced by the Misuse of Drugs (Amendment) Regulations 2003 (S.I. 1432) and the Misuse of Drugs (Amendment) (No. 3) Regulations 2005 (S.I. 3372) are described in Appendix 2. Schedule 5 was modified by the Misuse of Drugs and the Misuse of Drugs (Supply to Addicts) (Amendment) Regulations 2005 (S.I. 2864), which removed the exemption of preparations containing less than 0.1% cocaine (see Appendix 3).

The Misuse of Drugs (Amendment) (No. 2) Regulations 2005 gave exemption from a possession offence when mushrooms containing psilocin or an ester of psilocin were growing naturally and were not being cultivated, or were being picked for the purposes of destruction (Chapter 7).

A privately maintained list of all amendments to the Misuse of Drugs Regulations 2001 is available[4].

---
[4] http://www.rudifortson4law.co.uk/legaltexts/MISUSE~1.PDF

5.5 DRUGS ACT 2005

The Drugs Act 2005 had several objectives. One of these (Section 21) was intended to clarify the law regarding "magic mushrooms" (Chapter 7). But Section 2 was a much more controversial part of the Act. This created a new presumption of intent to supply where a defendant is found to be in possession of more than a certain quantity of controlled drugs. The controversy largely centred on what was tantamount to the idea that a defendant might be guilty of a certain offence unless proved otherwise. For Section 2 to operate, threshold amounts would have to be set for the main drugs of misuse. In the event, and following much consultation, it proved to be difficult to reach any consensus on what those thresholds should be. Arguments ranged from how the Courts would not be able to use their discretion and take other facts into consideration, fundamental legal issues raised by the reverse burden of proof, and the fact that amounts of drug consumed for personal use might vary both geographically and in time. It was also felt that the thresholds would be seen as acceptable levels for personal use, and would encourage drug dealers to carry just below the thresholds. No solution was proposed to the problem of how drug purity might be factored into the threshold amounts. In October 2006, the Government announced in a press release[5] that Section 2 of the Act would not be introduced for the present time. The remaining Sections of the Drugs Act were concerned with issues such as testing of arrestees for Class A drugs, and consequent assessment of those testing positive, providing additional powers to law-enforcement agencies to tackle dealers who swallow or hide drugs in body cavities, and requiring Courts to take account of aggravating factors, such as dealing near a school, when sentencing.

---

[5] http://press.homeoffice.gov.uk/press-releases/drugs-reclassification

CHAPTER 6
Generic Controls in the UK

6.1 "DESIGNER DRUGS"

The introduction in some countries of drug control involving generic or analogue definitions was largely driven by the appearance of "designer drugs". The original definition of designer drugs described them as *"Analogues, or chemical cousins, of controlled substances that are designed to produce effects similar to the controlled substances they mimic"*. The clear assumption is that a designer drug is not in itself a controlled substance. However, synthesis of designer drugs is as much driven by the availability of alternative precursor chemicals as the need to make a noncontrolled substance. Thus, a range of different precursors can be used to make a single drug and a single precursor can make several different drugs.

A few synthetic ring-substituted phenethylamines (*e.g.* STP; 2,5-dimethoxy-4-methylamphetamine and its bromine analogue DOB; bromo-STP; 4-bromo-2,5-dimethoxyamphetamine) had been subject to limited abuse in the US since the mid-1960s. In 1967, the first reports appeared of STP in the UK; the analysis of an illicit tablet, believed to have been imported from the US, was subsequently published[1]. By the mid-1970s other phenethylamines were reported in the US. This led to the inclusion in the Act, in 1977, of a generic definition (S.I. 1243). By the late 1980s, many more illicit phenethylamine designer drugs began to

---

[1] R.J. Lewis, D. Reed, A.G. Service and A.M. Langford, *The identification of 2-chloro-4,5-methylenedioxymethylamphetamine in an illicit drug seizure*, J. For. Sci., 2000, **45(5)**, 1119–1125.

appear. Most were produced in Europe, and by far the most common of these was MDMA (3,4-methylenedioxymethylamphetamine), but others such as 3,4-methylenedioxyamphetamine (MDA) and 3,4-methylenedioxyethylamphetamine (MDEA) were also found. Their attraction to youth culture is that they offer a mixture of stimulant and so-called empathogenic/entactogenic properties, and they are seen by users as safe drugs. Without exception, these ring-substituted phenethylamines had been well anticipated by the generic controls of 1977.

Substituted tryptamines had also been included in the Modification Order of 1977, but in the late 1970s other "designer drugs" began to appear. Two principal series were originally seen: those based on fentanyl and those structurally derived from pethidine (strictly α-prodine, the reverse ester of pethidine). The illicit production and abuse of fentanyls and pethidines was originally confined to the US. As narcotic analgesics, these substances offered similar effects to heroin, but in much smaller doses. Two of the substituted fentanyls seen at that time were α-methylfentanyl and 3-methylfentanyl; they are typically several hundred times more potent analgesics than morphine. Not surprisingly, the high potencies needed led to many accidental and fatal overdoses. The α-prodine series caused a notorious public health issue when it was found that a by-product of clandestine synthesis, known as MPTP (1-methyl-4-phenyl-1,2,5,6-tetrahydropyridine), produced a rapid and irreversible chemically induced Parkinson's disease. Since then clandestine chemists have shown little further interest in pethidine/α-prodine derivatives. However, interest in the fentanyl family continues to arise sporadically. Thus, 350 fatal poisonings caused by illicit fentanyl were reported[2] in Illinois in the period 2005–2007. In Estonia, a recent epidemic of fatal poisonings[3] was caused by 3-methylfentanyl. There have also been isolated reports of the appearance of 4-fluorofentanyl in Europe. Many pethidine and fentanyl derivatives were brought within the scope of the Act by generic controls in 1986 (S.I. 2230).

Derivatives of the plant-based drugs, particularly cannabinols and cocaine have not been widely evaluated by clandestine chemists. This may partly reflect the wide availability of the parent product, but also because they are more complex structures than, for example, phenethylamines or tryptamines. However, that may change following the discovery of an illicit synthetic cocaine laboratory in Spain in 2001. If the means are available to synthesise cocaine, then it would be particularly

---

[2] J.S. Denton, E.R. Donoghue, J. McReynolds and M.B. Kalelkar, *An epidemic of illicit fentanyl deaths in Cook County, Illinois: September 2005 through April 2007*, J. For. Sci., 2008, **53 (2)**, 452–454
[3] I. Ojanperä, M. Gergov, M. Liiv, A. Riikoja and E. Vuori, *An epidemic of fatal 3-methylfentanyl poisoning in Estonia*, Int. J. Legal Med., 2008, **122**, 115–121

easy to create a series with varied substitution patterns in the phenyl ring. As mentioned later, the only known illicit derivative of a cannabinol is the acetyl ester of tetrahydrocannabinol.

In the following sections, the generic controls based on salts, ethers and esters are first described. These extensions had their origins in the UN Conventions. The control of stereoisomers is then covered. Although more fully developed in the Misuse of Drugs Act, some control of stereoisomers was already in place in UK drugs legislation before 1971. In subsequent sections, the generic classification of anabolic steroids, barbiturates, cannabinols, ecgonine derivatives, fentanyls, lysergamide derivatives, pentavalent derivatives of morphine, pethidines, phenethylamines and tryptamines are described. The generic definition of ring-substituted phenethylamines has proved to be remarkably far-sighted. Although far fewer "designer" tryptamines have been found, the generic controls have again been useful. However, the definitions of other groups have hardly been tested.

In Chapter 10, the New Zealand and US approaches to generic control and analogue legislation are discussed. But in most countries, drugs legislation closely follows the item by item listing of substances in the 1961 and 1971 UN Conventions. Despite the UK having over thirty years of experience of operating generic controls, numerous arguments against them or perceived difficulties continue to be raised. These include:

- **They would hinder the development in the pharmaceutical industry of novel compounds for legitimate clinical use.** This has not been a problem in the UK. Even if the pharmaceutical industry did wish to develop substances that were covered by generic controls, it would be a simple matter to either issue licences or modify the legislation.
- **Control of chemical groups may cover substances with a range of different pharmacological effects and some with no effects whatsoever.** Because the Act relies on the concept of actual or potential social harm, rather than the specific pharmacological or toxicological properties of a controlled drug, no great difficulty arises from the introduction of generic control. This is a valid argument in those jurisdictions (and the UN itself) where there is an *a priori* need to review the pharmacological and toxicological properties of every substance considered for control. It is quite certain that amongst the essentially infinite number of generically defined substances there will be compounds that have little abuse potential and some may have no physiological effect of any sort. Without these effects, a substance will not be marketed by the pharmaceutical industry and neither will it be produced as a misusable drug.

- **Useful medicines and other substances will be inadvertently controlled.** Provided that the definitions of included substances are sufficiently rigorous, this should rarely happen. In the generic definition of phenethylamines (see later), a specific exclusion was made for the (now obsolete) active pharmaceutical ingredient methoxyphenamine.
- **Generic controls will be difficult to comprehend.** One of the most complex definitions in the Misuse of Drugs Act involves ring-substituted phenethylamines, but in the past thirty years many tens of thousands of witness statements, involving the identification of MDMA in seized samples, have been submitted by UK forensic science laboratories. These statements have incorporated the definition without any apparent problems.

The phenomenon of designer drugs is not restricted to controlled drugs. A recent example was the appearance, in so-called "herbal aphrodisiacs", of various unlicensed analogues of sildenafil[4].

6.2 SALTS

The salts of all controlled drugs are controlled to the same degree as the parent. A salt is the product of reacting a base with an acid. Like many physiologically active chemicals, controlled drugs are mostly bases, often described as nitrogenous bases or, in some cases, alkaloids. For various reasons, including stability and ease of handling, the salts, especially hydrochlorides and sulfates, less commonly tartrates and phosphates, are more often seen in both commercial and illicit products than the parent substances. Structures (6.1) and (6.2) show two examples of the formation of salts.

Structure (6.1) The reaction of amphetamine with sulfuric acid to form the sulfate salt

---

[4] B.J. Venhuis, L. Blok-Tip and D. de Kaste, *Designer drugs in herbal aphrodisiacs,* For. Sci. Int., 2008, **177**, e-25–e-27

Structure (6.2) The reaction of methylamphetamine with hydrochloric acid to form the hydrochloride salt

Acidic drugs are uncommon; the best examples amongst controlled drugs are the barbiturates. Reaction of a 5,5-disubstituted barbituric acid with sodium hydroxide (a base) produces the sodium salt (Structure (6.3)).

Structure (6.3) The reaction of a 5,5-disubstituted barbituric acid with sodium hydroxide to form the disodium salt

Unless a prosecution wishes to bring a charge of production of a base from a salt (*e.g.* crack cocaine from cocaine hydrochloride), then it is not necessary for the forensic chemist to identify whether a questioned substance is in its free form (base or acid) or a particular salt.

6.3 ESTERS AND/OR ETHERS

The esters or ethers of Class A substances and of the Class C anabolic/androgenic steroids are subject to the same controls as their unmodified parents, unless that ester or ether is already specified elsewhere in Schedule II. Only structures with a hydroxyl (-OH), sulfydryl (-SH) or a suitable acid (*e.g.* carboxylic -COOH) group commonly form esters, and only hydroxyl and sulfydryl groups form ethers. Amongst those basic drugs listed in the Act, which are able to form an ester or an ether, only the hydroxyl function is found. Schedule 2 of the Misuse of Drugs Act refers specifically to control of *"Any ester or ether . . . ".* This was deliberately designed to be more inclusive. In other words, a substance that is both an ester *and* an ether is not controlled. An example here is thebacon, which, as an ester and an ether of hydromorphone, is listed by name in Part 1 of Schedule 2.

6.3.1 Esters

An example of ester formation is the conversion of morphine to diamorphine (the diacetyl ester of morphine) as shown in Structure (6.4). This process is used, for example, in the illicit production of heroin (crude diamorphine). Diamorphine slowly hydrolyses in damp conditions or rapidly in aqueous alkaline solutions to produce 6-*O*-monoacetylmorphine (Structure (6.5)). Monoacetylmorphine is still an ester of morphine and therefore remains a Class A controlled drug.

Structure (6.4) The esterification of morphine to diamorphine (Ac$_2$O is acetic anhydride)

Structure (6.5) The hydrolysis of diamorphine to form 6-*O*-monoacetylmorphine

Another example of an ester is psilocybin, the naturally occurring phosphate of psilocin (Structure (6.6)). Both psilocin and psilocybin are found in certain fungi of the *Psilocybe* genus (so-called magic mushrooms; Chapter 7).

Structure (6.6) Psilocybin, the naturally occurring phosphate ester of psilocin

The illicit production of the acetyl ester of tetrahydrocannabinol (THC) has been recorded in clandestine laboratories. It is claimed that the resulting THC acetate is a more potent drug.

Esters of the Class C steroids are quite common in commercial formulations. Structures (6.7) and (6.8) show testosterone and its propionate ester, respectively. Other common esters of testosterone are the 17β-cyclopentanepropionate (cypionate) and the 17β-undecanoate (undecyclate).

Structure (6.7) Testosterone, an androgenic anabolic steroid

Structure (6.8) The propionate ester of testosterone

6.3.2 Ethers

A number of controlled drugs, principally certain opioids, are ethers. Neither codeine (the 3-methyl ether of morphine; Structure (6.9)), dihydrocodeine (the 3-methyl ether of dihydromorphine; Structure (6.10)), pholcodine (the 3-[morpholinoethyl] ether of morphine;

Structure (6.11)) nor ethylmorphine (the 3-ethyl ether of morphine; Structure (6.12)) are Class A drugs because they are already listed under Class B. However, other ethers of morphine or dihydromorphine or of another Class A drug would be controlled under Class A. Various substituted tryptamines, which would qualify as Class A esters or ethers are described in Appendix 20.

Structure (6.9) Codeine, the 3-methyl ether of morphine

Structure (6.10) Dihydrocodeine (the 3-methyl ether of dihydromorphine)

Structure (6.11) Pholcodine (the 3-[morpholinoethyl] ether of morphine)

Structure (6.12) Ethylmorphine (the 3-ethyl ether of morphine)

6.4 STEREOISOMERISM

Stereoisomers are substances with the same molecular formula, but with different spatial arrangements of their atoms in the molecule, leading to different physical and pharmacological properties. They have to contain in their molecule one or more so-called chiral centres. Different spatial arrangements at one chiral centre give rise to molecules that are related as mirror images, and are called enantiomers. Such molecules are normally optically active (they rotate the plane of polarised light) and were formerly designated *(d)* [from *dexter*] or (+) or were designated *(l)* [from *laevus*] or (−). The (+) and (−) forms cause rotation to the right and left, respectively. They may also form an optically neutral racemate, *i.e.* a mixture of equal numbers of (+) and (−) molecules, shown as *(dl)* or "(±)". However, these terms are now obsolete; the present standard designation for steric configuration (the Cahn–Ingold–Prelog or CIP rule) is the *R* (rectus, right) and *S* (sinister, left) notation. It is related to the absolute steric configuration of substituents at chiral centres. The CIP rule replaced an earlier absolute system that used the designation D (derived from **D**extrose) and L (derived from **L**aevulose, *i.e.* fructose). Neither the CIP system (*R* or *S*) nor the D/L systems bear any correlation with the direction of optical rotation.

Stereoisomerism is common amongst controlled drugs. In all cases, the chiral centre involves an asymmetric carbon atom, that is to say one having four different substituents. With a few named exceptions, discussed later, all stereoisomers of controlled drugs are controlled. As with salts, in a criminal trial there is no need for the prosecution to name a particular stereoisomeric form. Structure (6.13) shows the two enantiomers of amphetamine, where each is a mirror image of the other such that neither is superimposable on the other.

Structure (6.13) The two enantiomers of amphetamine

Whereas enantiomeric pairs are common, there are far fewer instances of controlled drugs with two or more chiral centres. A good example occurs in 1-hydroxy-1-phenyl-2-aminopropane. Here, there are two asymmetric carbon atoms giving rise to four stereoisomers as set out in Structure (6.14). The 1*S*,2*S* enantiomer, also known as (+)-norpseudoephedrine, is cathine (Class C). This is a mirror image of its 1*R*,2*R* enantiomer, *i.e.* (−)-norpseudoephedrine. The 1*S*,2*R* and 1*R*,2*S* enantiomers are known as (+)- and (−)-norephedrine, respectively. Each of the four stereoisomers is optically active; aqueous solutions of enantiomers at the same concentration will rotate the plane of polarised light to an equal extent, but in opposite directions. However, opposite pairs, *e.g.* (+)-norephedrine and (+)-norpseudoephedrine, known as diastereoisomers, will not produce an equal rotation. Thus, in this example, there are two pairs of enantiomers and four pairs of diastereoisomers.

The two stereoisomers of norephedrine, when present as a racemic mixture, are known as phenylpropanolamine, although, confusingly, this term has sometimes been used to refer only to the 1*R*, 2*S* isomer. Cathine can be distinguished from its noncontrolled diastereoisomers, *i.e.* (+)- and (−)-norephedrine (phenylpropanolamine) using thin-layer chromatography. Phenylpropanolamine, as a useful decongestant drug, is excluded from control by paragraph 2 of Part III of Schedule 2 to the Act. Although it has been withdrawn from the market in the US because of fears that it increases the risk of haemorrhagic stroke, phenylpropanolamine is still available in the UK. Norephedrine can be used as a precursor to amphetamine; it was recently added to Table I of UN1988 and the corresponding EU and UK legislation (Appendix 5). A less common illicit use of norephedrine is as a precursor to the stimulant drug 4-methylaminorex[5].

Ephedrine and pseudoephedrine form an exactly similar group of four stereoisomers. An extreme case of stereoisomeric complexity amongst controlled drugs occurs with pentazocine, which has three chiral centres and therefore three pairs of enantiomers and three racemates.

(+)-Norpseudoephedrine (1*S*, 2*S*)   (−)-Norpseudoephedrine (1*R*, 2*R*)

(+)-Norephedrine (1*S*, 2*R*)   (−)-Norephedrine (1*R*, 2*S*)

Structure (6.14) The four stereoisomers of 1-hydroxy-1-phenyl-2-aminopropane

---

[5] W.R. Rodriguez and A.A. Russell A. *Synthesis of trans-4-methylaminorex from norephedrine and potassium cyanate*, Microgram Journal, 2005, **3(3–4)** http://www.dea.gov/programs/forensicsci/microgram/journal_v3_num34/journal_v3_num34_pg6.html

Further exceptions for certain stereoisomers are made in the Act for dextromethorphan and dextrorphan. These are both of clinical value although dextromethorphan is occasionally misused. Their enantiomers (*i.e.* levomethorphan and levorphanol, respectively) have much greater abuse potential and are both Class A controlled drugs.

In 1998, following a proposal from the Spanish Government, the WHO considered extending control of substances listed in UN1971 to isomers, esters, ethers and "analogues". Although some of these proposals would have brought the 1971 Convention into line with current UK practice, the proposals were rejected. Despite the positive experience of the UK with generic controls, the WHO considered that the changes might have a negative impact on legitimate industry. It also stated that control of "analogues" would contradict its mandate of evaluating individual substances. The proposed control of isomers, as opposed to *stereoisomers* in the Misuse of Drugs Act, was widely regarded as being too vague.

6.5 ANABOLIC STEROIDS

In Part III of Schedule 2, paragraph 1(c) defines Class C anabolic/androgenic steroids as:

"any compound (not being Trilostane or a compound for the time being specified in subparagraph (b) above) structurally derived from 17-hydroxyandrostan-3-one or from 17-hydroxyestran-3-one by modification in any of the following ways, that is to say,

  (i) *by further substitution at position 17 by a methyl or ethyl group;*
 (ii) *by substitution to any extent at one or more of the positions 1,2,4,6,7,9,11 or 16, but at no other position;*
(iii) *by unsaturation in the carbocyclic ring system to any extent, provided that there are no more than two ethylenic bonds in any one carbocyclic ring;*
(iv) *by fusion of ring A with a heterocyclic system"*

Structure (6.15) shows the basic steroid ring-numbering system. Structures (6.16) and (6.17) show the general structure of the two steroids (*i.e.* 17-hydroxyandrostan-3-one and 17-hydroxyestran-3-one) upon which the above rules operate. Testosterone, a substance listed specifically is shown in Structure (6.18). None of the four further anabolic steroids added to the Act in 2003 as named Class C drugs (Chapter 9) was covered by the above definition. Trilostane (Structure (6.19)), which would otherwise be included in the generic definition, is specifically excluded since it has clinical value as an adrenocortical suppressant used in the treatment of breast cancer. Structure (6.20) shows 4-androstene-3,17-dione: a steroid that fails the generic test because the substitution at position 17 is not by a methyl or ethyl group. It was added specifically to the Act in 2003.

Structure (6.15) The ring-numbering system in steroids

Structure (6.16) 17-Hydroxyandrostan-3–one

Structure (6.17) 17-Hydroxyestran-3–one

Structure (6.18) Testosterone, a steroid listed specifically in the Act

Structure (6.19) Trilostane, specifically excluded from control

Structure (6.20) 4-Androstene-3,17-dione, a steroid not covered by the generic definition, but added to the Act in 2003

Since most countries take their lead on drug control from the UN Conventions, it is not surprising that few have controlled anabolic steroids. These substances are neither narcotics nor psychotropics and fall beyond the current scope of the international treaties. In the US Controlled Substances Act, 59 anabolic substances are named. In both countries, esters or ethers of the controlled substances are subsumed, but only the UK has extended the list to include generic definitions as well as related products such as growth hormones, clenbuterol (a β2-adrenergic agonist) and nonsteroidal anabolic agents (zeranol and zilpaterol). The full US list can be found at the website of the DEA Office of Diversion Control[6].

6.6 BARBITURATES

In Part II of Schedule 2, paragraph 1(c) defines Class B barbiturates as: *"any 5,5 disubstituted barbituric acid"*. Structure (6.21) shows a 5,5 disubstituted barbituric acid. In practice, all clinically useful barbiturates have $R^{5\alpha}$ and $R^{5\beta}$=alkyl, alkenyl or aryl. The *N*-substituted barbiturate, methylphenobarbitone, does not comply with this rule (because of the *N*-methyl substituent) and is therefore listed specifically as a Class B drug. *"Barbituric acid"* means oxo-barbituric acid so thiobarbiturates

---
[6] http://www.deadiversion.usdoj.gov/21cfr/cfr/1300/1300_01.htm

Table 6.1 Twelve barbiturates[a] listed in the UN Convention on Psychotropic Substances 1971 – see Structure (6.21)

Name	R^1	$R^{5\alpha}$	$R^{5\beta}$
Allobarbital	H	Allyl	Allyl
Amobarbital	H	Ethyl	iso-Pentyl
Barbital	H	Ethyl	Ethyl
Butalbital	H	Allyl	iso-Butyl
Butobarbital	H	Ethyl	n-Butyl
Cyclobarbital	H	Ethyl	1-Cyclohexen-1-yl
Methylphenobarbital	Methyl	Ethyl	Phenyl
Pentobarbital	H	Ethyl	1-Methylbutyl
Phenobarbital	H	Ethyl	Phenyl
Secbutabarbital	H	Ethyl	s-Butyl
Secobarbital	H	Allyl	1-Methylbutyl
Vinylbital	H	Vinyl	1-Methylbutyl

[a]Secobarbital is listed in Schedule II, amobarbital, butalbital, cyclobarbital and pentobarbital are in Schedule III; the remainder are in Schedule IV

such as thiopentone (Structure (6.22)) are not controlled, while barbituric acid itself, having no 5,5-disubstitution, is likewise excluded.

The twelve barbiturates listed in UN1971 are shown in Table 6.1; this illustrates the economy of the UK generic definition. The generic definition of 5,5 disubstituted barbituric acids enables an unknown substance to be uniquely assigned to this group by measuring its UV absorption spectra at pH values corresponding to the formation of a di-anion, a mono-anion and a neutral species.

Structure (6.21) A 5,5-disubstituted barbituric acid

Structure (6.22) Thiopentone

Most barbiturates are covered by Schedule 3 of the Regulations, but quinalbarbitone (secobarbital), is named specifically in Schedule 2 of the Regulations. This is partly because of its higher intrinsic toxicity. Barbitone (5,5-diethylbarbituric acid), a substance often used in buffering solutions, clearly qualifies as a Class B drug. But the Regulations exempt *"a person in charge of a laboratory when acting in his capacity as such"* from the restrictions on possession and supply of barbitone (and any other Schedule 3 drug) when in the form of buffering solutions. The definition of a 5,5-disubstituted barbituric acid also captures substances that are not listed in UN1971 (*e.g.* 5,5-diphenylbarbituric acid). However, barbiturates are now rarely used therapeutically.

6.7 CANNABINOLS

The main psychoactive principle in cannabis is Δ^9-tetrahydrocannabinol (THC). In Structure (6.23), the substituent R at position 3 is pentyl. The unsaturated bond in the cyclohexene ring is located between C_9 and C_{10} in the more common dibenzopyran ring-numbering system. In the monoterpenoid ring-numbering system, that double bond is between C_1 and C_2. The naturally occurring active isomer Δ^8-THC, where the unsaturated bond in the cyclohexene ring is located between C_8 and C_9, is found in much smaller amounts. Although Δ^9-tetrahydrocannabinol can exist in four stereoisomeric forms, only the *R,R-trans*-stereoisomer, a substance known by its INN as dronabinol, occurs naturally.

Structure (6.23) Δ^9-Tetrahydrocannabinol showing ring-numbering in the dibenzopyran system (left) and the partial ring-numbering in the monoterpenoid system (right). R = C_5H_{11}

Two precursor substances, Δ^9-tetrahydrocannabinol-2-oic acid and Δ^9-tetrahydrocannabinol-4-oic acid (THCA) are also present in cannabis, sometimes in large amounts. During smoking, THCA is converted to THC, although other substances are also formed. Quantitative analysis of THC in cannabis is usually achieved by gas-chromatography; the precursor acids are pyrolysed in the injection port thereby enabling total THC to be estimated. The concentration of THC in imported herbal cannabis and resin is typically 5%, but may be much higher in cannabis grown under intensive conditions (Chapter 7). Other major components in cannabis are cannabinol (CBN; an oxidation product of THC; Structure (6.24), where the substituent R at position 3 is pentyl) and cannabidiol (CBD; Structure (6.25)). Some CBD, like THC, occurs naturally as precursor cannabidiolic acids. Cannabinol, except where contained in cannabis or cannabis resin, and cannabinol derivatives were once Class A drugs. As with cannabis and cannabis resin, cannabinol and cannabinol derivatives were moved to Class C in 2004. In early 2008, the ACMD recommended that, these "cannabis products" should stay in Class C, but the Government ignored this advice and decided that they would be placed in Class B (Chapter 11).

In UN1971, six specific isomers and stereochemical variants of THC are listed by name together with two related substances (DMHP and parahexyl) as shown in Table 6.2. However, the generic clause in Part IV of Schedule 2 of the Misuse of Drugs Act defines *"cannabinol derivatives"* as *"...the following substances, except where contained in cannabis or cannabis resin, namely tetrahydro derivatives of cannabinol and 3-alkyl homologues of cannabinol or of its tetrahydro derivatives"*. A controlled drug arises if, in Structures (6.23) or (6.24), the substituent R at position 3 is any alkyl group with 5 or more carbon atoms (see Chapter 3). This definition excludes, for example, CBD, which is not a simple derivative of cannabinol or tetrahydrocannabinol. Also excluded are the natural carboxylic acid precursors (THCA), one of which is shown in Structure (6.26), and the metabolite 11-hydroxy-Δ^9-THC. Like cannabis and cannabis resin, controlled cannabinols are included in Schedule 1 of the Misuse of Drugs Regulations 2001. Other noncontrolled compounds present in cannabis and cannabis resin

Table 6.2 Six isomers and stereochemical variants of THC and two related substances (DMHP and Parahexyl) as listed in the UN Convention on Psychotropic Substances 1971.

7,8,9,10-Tetrahydro-6,6,9-trimethyl-3-pentyl-6*H*-dibenzo[*b,d*]pyran-1-ol
(9*R*,10a*R*)-8,9,10,10a-Tetrahydro-6,6,9-trimethyl-3-pentyl-6*H*-dibenzo[*b,d*]pyran-1-ol
(6a*R*,9*R*,10a*R*)-6a,9,10,10a-Tetrahydro-6,6,9-trimethyl-3-pentyl-6*H*-dibenzo[*b,d*]pyran-1-ol
(6a*R*,10a*R*)-6a,7,10,10a-Tetrahydro-6,6,9-trimethyl-3-pentyl-6*H*-dibenzo[*b,d*]pyran-1-ol
6a,7,8,9-Tetrahydro-6,6,9-trimethyl-3-pentyl-6*H*-dibenzo[*b,d*]pyran-1-ol
(6a*R*,10a*R*)-6a,7,8,9,10,10a-Hexahydro-6,6-dimethyl-9-methylene3-pentyl-6*H*-dibenzo[*b,d*]pyran-1-ol
3-(1,2-Dimethylheptyl)-7,8,9,10-tetrahydro-6,6,9-trimethyl-6*H*-dibenzo[*b,d*]pyran-1-ol (DMHP)
3-Hexyl-7,8,9,10-tetrahydro-6,6,9-trimethyl-6*H*-dibenzo[*b,d*]pyran-1-ol (Parahexyl)

Generic Controls in the UK

include the cannabivarins and cannabichromenes; together with THC, CBN and CBD, they are collectively known as cannabinoids.

The inclusion of cannabinol in the Misuse of Drugs Act (see Tables 5.1 and A15.1) is an anomaly; it is not psychoactive, not an obvious precursor to THC and is not listed in UN1971.

Structure (6.24) Cannabinol (R = C_5H_{11})

Structure (6.25) Cannabidiol

Structure (6.26) Tetrahydrocannabinol-2-oic acid

6.8 ECGONINE DERIVATIVES

Ecgonine (3-hydroxy-8-methyl-8-azabicyclo[3.2.1]octane-2-carboxylic acid) is listed as a Class A drug in paragraph 1(a) of Part I of Schedule 2 to the Act. It is not an abusable substance *per se*, but one of several drug intermediates (*i.e.* precursors) that occur in the Act (Appendix 4). The full entry reads *"Ecgonine, and any derivative of ecgonine which is convertible to ecgonine or to cocaine"*. The relationship between ecgonine and cocaine is shown in Structures (6.27) and (6.28). It will be seen that they differ at both the 2- and 3-positions of the tropane ring.

Structure (6.27) Ecgonine showing the ring-numbering system

Structure (6.28) Cocaine

Bearing in mind the definition of a derivative given in Chapter 3, Structure (6.29) shows benzoylecgonine: a substance that would qualify as a controlled derivative of ecgonine because it can be converted to cocaine by esterification and converted to ecgonine by hydrolysis. Structure (6.30) shows a substance that had been under development as a potential anti-depressant drug. Although the synthesis of this compound starts from cocaine, it does not qualify as a controlled drug because it cannot be converted to either ecgonine or cocaine in a single stage.

Structure (6.29) A controlled derivative of ecgonine (benzoylecgonine)

Structure (6.30) A noncontrolled derivative of ecgonine

6.9 FENTANYLS

In Part I of Schedule 2, paragraph 1(d) defines Class A fentanyls as:

"*any compound (not being a compound for the time being specified in subparagraph (a) above) structurally derived from fentanyl by modification in any of the following ways, that is to say,*

(i) *by replacement of the phenyl portion of the phenethyl group by any heteromonocycle whether or not further substituted in the heterocycle;*
(ii) *by substitution in the phenethyl group with alkyl, alkenyl, alkoxy, hydroxy, halogeno, haloalkyl, amino or nitro groups;*
(iii) *by substitution in the piperidine ring with alkyl or alkenyl groups;*
(iv) *by substitution in the aniline ring with alkyl, alkoxy, alkylenedioxy, halogeno or haloalkyl groups;*
(v) *by substitution at the 4-position of the piperidine ring with any alkoxycarbonyl or alkoxyalkyl or acyloxy group;*
(vi) *by replacement of the N-propionyl group by another acyl group*"

Structure (6.31) shows fentanyl upon which the above rules operate. There are four named fentanyl derivatives in Class A (alfentanil, carfentanil, lofentanil and sufentanil). All are covered by the above definition; they were added to Part I of Schedule 2 in 1988 at the same time as the above generic definition was introduced. As examples of controlled fentanyl derivatives, lofentanil (Structure (6.32)) passes the generic test because the 3-methyl substituent in the piperidine ring and the 4-methoxycarbonyl substituent are both permitted. In sufentanil (Structure (6.33)), the replacement of the phenethyl group by thiophenylethyl as well as the 4-methoxymethyl substituent in the piperidine ring are both permitted. Remifentanil (Structure (6.34)) fails to meet the fentanyl derivatives generic description because the phenethyl group has been replaced with a group not specified in the definition. Remifentanil has been added to the Act as a named Class A drug (Chapter 5).

The two most commonly reported illicit fentanyl derivatives (α-methylfentanyl[7] and 3-methylfentanyl) are shown in Structures (6.35) and (6.36), respectively. They are both captured by the generic definition.

---

[7] http://en.wikipedia.org/wiki/Alphamethylfentanyl

Structure (6.31) Fentanyl

Structure (6.32) Lofentanil: a Class A substance controlled by the generic definition of a substituted fentanyl

Structure (6.33) Sufentanil: a Class A substance controlled by the generic definition of a substituted fentanyl

Structure (6.34) Remifentanil: a Class A substance that is not covered by the generic definition of a substituted fentanyl, but is listed specifically

Structure (6.35) α-Methylfentanyl: an illicit product and a Class A substance captured by the generic definition of a substituted fentanyl

Structure (6.36) 3-Methylfentanyl: an illicit product and a Class A substance captured by the generic definition of a substituted fentanyl

6.10 LYSERGIDE AND DERIVATIVES OF LYSERGAMIDE

Structure (6.37) shows lysergamide with all nitrogen substituents identified. Lysergamide is not listed in the UN Conventions, but is a Class A controlled drug in the UK. Lysergide (LSD) is named specifically as a Class A drug; it is the diethylamide of lysergic acid where the ethyl groups are at R' and R". Generic control is extended to *"Lysergide and other N-alkyl derivatives of lysergamide"*. A Class A drug therefore arises if R' and/or R" and/or R^1 is alkyl, regardless of whether any other substituents are present. Table 6.3 shows four derivatives of lysergamide listed in TIHKAL (see Bibliography) with their control status under the Act.

Structure (6.37) Lysergamide (R' = R" = R^1 = H; R^6 = methyl) showing substitution patterns at the three nitrogen atoms

It might be questioned whether lysergamide alkylated at R' and/or R" and where R^6 is an alkyl group other than methyl, would be controlled. This is a "derivative of a derivative" argument. If R^6 is not methyl, then

Table 6.3 Derivatives of lysergamide[a] listed in TIHKAL and their control status under the Misuse of Drugs Act – see Structure (6.37)

TIHKAL Ref.	R'	R"	R^1	R^6	Controlled?
1	Ethyl	Ethyl	H	Allyl	No
12	Ethyl	Ethyl	H	Ethyl	No
26	Ethyl	Ethyl	H	Methyl	Yes (Lysergide)
51	Ethyl	Ethyl	H	Propyl	No
Lysergamide	H	H	H	Methyl	Yes

[a]Lysergamide itself is not listed in TIHKAL

the core structure ought no longer to be regarded as "lysergamide", and none of its *N*-alkyl derivatives would fall to control. It is believed that the original legislation was not intended to control 1-alkyl lysergamide alkanolamide derivatives such as methysergide, a drug used to treat migraine, for which R^1=methyl, R^6=methyl, R'=CH(CH$_2$-OH)CH$_2$CH$_3$ and R''=H.

6.11 PENTAVALENT DERIVATIVES OF MORPHINE

In Part I of Schedule 2, control is extended to *"morphine methobromide, morphine N-oxide and other pentavalent nitrogen morphine derivatives"*. However, these substances are rarely used or misused. Structure (6.38) shows morphine *N*-oxide.

Structure (6.38) Morphine *N*-oxide: a pentavalent derivative of morphine

6.12 PETHIDINES

In Part I of Schedule 2, paragraph 1(e) defines Class A pethidines as:

"any compound (not being a compound for the time being specified in subparagraph (a) above) structurally derived from pethidine by modification in any of the following ways, that is to say,

(i) *by replacement of the 1-methyl group by an acyl, alkyl whether or not unsaturated, benzyl or phenethyl group, whether or not further substituted;*
(ii) *by substitution in the piperidine ring with alkyl or alkenyl groups or with a propano bridge, whether or not further substituted;*
(iii) *by substitution in the 4-phenyl ring with alkyl, alkoxy, aryloxy, halogeno or haloalkyl groups;*
(iv) *by replacement of the 4-ethoxycarbonyl by any other alkoxycarbonyl or any alkoxyalkyl or acyloxy group;*
(v) *by formation of an N-oxide or of a quaternary base"*

Generic Controls in the UK 67

Structure (6.39) shows pethidine (meperidine) upon which the above rules operate. Several Class A drugs are closely related to pethidine. Three pethidine intermediates are listed by name in the Act; none complies with the above definition and all three had originally been listed in the Dangerous Drugs Act 1964, many years before the generic definition was introduced. They are 4-cyano-1-methyl-4-phenylpiperidine, 1-methyl-4-phenylpiperidine-4-carboxylic acid and 4-phenylpiperidine-4-carboxylic acid ethyl ester (Appendix 4, Table A4.1). In the case of 1-methyl-4-phenylpiperidine-4-carboxylic acid (Structure (6.40)), it fails the test because there is a 4-carboxylic acid moiety instead of a 4-ethoxycarbonyl group on the piperidine ring. Several other substances fail the above rules including allylprodine (Structure (6.41)), alphameprodine, alphaprodine, difenoxin, diphenoxylate, hydroxypethidine and phenoperidine.

Structure (6.39) Pethidine

Structure (6.40) 1-methyl-4-phenylpiperidine-4-carboxylic acid (also known as Pethidine intermediate C), which is not covered by the generic definition of a substituted pethidine, but listed by name as a Class A drug

Structure (6.41) Allylprodine: a Class A substance not controlled by the generic definition of a substituted pethidine

6.13 PHENETHYLAMINES

In Part I of Schedule 2, paragraph 1(c) defines Class A phenethylamines as:

"any compound (not being methoxyphenamine or a compound for the time being specified in subparagraph (a) above) structurally derived from phenethylamine, an N-alkylphenethylamine, α-methylphenethylamine, an N-alkyl-α-ethyl-phenethylamine, α-ethylphenethylamine, or an N-alkyl-α-ethylphenethylamine by substitution in the ring to any extent with alkyl, alkoxy, alkylenedioxy or halide substituents, whether or not further substituted in the ring by one or more other univalent substituents".

Structure (6.42) shows phenethylamine upon which the above rules operate.

Structure (6.42) Phenethylamine (β-phenylethylamine) showing substitution patterns

To qualify as a Class A drug, the following criteria must be satisfied:
$R' = H$ or alkyl
$R'' = R^{\alpha 1} = R^{\beta 1} = R^{\beta 2} = H$
$R^{\alpha 2} = H$, methyl or ethyl
R^2 = alkyl, alkoxy, alkylenedioxy or halogen (either singly or in any combination) with or without any other substitution in the ring

The focus of this rather daunting definition is ring-substitution in amphetamine-like molecules. The reasoning behind this is that the attachment of other atoms (especially oxygen, sulfur or halogen) to one or more of the carbon atoms (commonly the 2-, 4- or 5-positions) in the aromatic ring of phenethylamine leads to major changes in pharmacological properties. Whilst amphetamine and many of its side-chain isomers and other simple derivatives (*e.g.* methylamphetamine, methcathinone, benzphetamine) are all central nervous system stimulants, suitable substitution in the ring can create hallucinogens (*e.g.* mescaline) or empathogenic/entactogenic agents that may or may not retain some stimulant activity. The well-known controlled drug MDMA, a member of the ecstasy group of so-called entactogenic stimulants, is more formally known as 3,4-methylenedioxymethylamphetamine or, fully systematically, as N-methyl-3,4-methylenedioxyphenylpropan-2-amine. Yet none of these terms nor any other direct synonym will be found in the Act.

The specific exception of methoxyphenamine (Structure (6.47)) was made because this drug, a prescription bronchodilator in the proprietary product Orthoxine, would have fallen to control under the subsequent definition. However, methoxyphenamine was withdrawn from general use in the UK in 1986 and its continued mention in the legislation is now redundant. The generic definition deliberately excludes from control ring-hydroxyl phenethylamines, a group that includes naturally occurring products such as dopamine, tyramine and adrenaline as well as clinically useful substances such as 4-hydroxyamphetamine. However, substances with ring hydroxyl substitution *and* one or more of the specified generic substituents would qualify for control.

Structures (6.43) to (6.50) show examples of phenethylamines, which either do or do not represent controlled drugs under the generic definition. Structure (6.43) depicts the well-known example of MDMA (3,4-methylenedioxy-methylamphetamine) which qualifies as a Class A controlled drug.

Structure (6.43) MDMA (3,4-methylenedioxymethylamphetamine): a Class A substance controlled by the generic definition of a substituted phenethylamine

The substance in Structure (6.44), one of the "PIHKAL" drugs, fails the generic test because of the β-methoxy substituent, but is listed by name in paragraph 1(ba) of Part I of Schedule 2 as 4-bromo-β,2,5-trimethoxyphenethylamine.

Structure (6.44) A phenethylamine derivative from the 'PIHKAL' list (4-bromo-β,2,5-trimethoxyphenethylamine), not covered by the generic definition of a substituted phenethylamine, but listed by name as a Class A drug

Structure (6.45) shows an example of a noncontrolled phenethylamine (N,α-dimethyl-4-nitrophenethylamine); it has been described as having pharmacological properties similar to those of analogous phenethylamines[8]. But it is not listed specifically and fails the generic test because of the 4-nitro substituent and the absence of other defined ring-substituents.

Structure (6.45) A noncontrolled phenethylamine (N, α-dimethyl-4-nitrophenethylamine), not covered by the generic definition of a substituted phenethylamine and not listed by name in the Act

Structure (6.46) shows mebeverine, an anti-spasmodic drug, which does not qualify as a controlled phenethylamine. It too is not listed specifically and does not get captured by the generic test because of the complex substituent on the nitrogen atom.

Structure (6.46) Mebeverine, a noncontrolled phenethylamine not covered by the generic definition of a substituted phenethylamine and not listed by name in the Act

Structure (6.47) shows methoxyphenamine (o-methoxymethylamphetamine) that is specifically excluded from control.

Structure (6.47) Methoxyphenamine, specifically excluded from control

Structure (6.48) shows the *p*-isomer (PMMA) of methoxyphenamine: a substance that falls within the generic definition. PMMA (paramethoxymethylamphetamine; methyl-MA) has been seen in drug seizures in Europe and was subjected to risk assessment by the EMCDDA in late

---

[8] J. Knoll, E.S. Vizi and Z. Ecseri, *Psychomimetic methylamphetamine derivatives*, Arch. Int. Pharmacodyn., 1966, **159(2)**, 442–451

2001 (Chapter 4). Although their mass spectra, for example, do show some small differences, the forensic analyst would need to ensure that methoxyphenamine and PMMA could be clearly distinguished from each other.

Structure (6.48) Paramethoxymethylamphetamine (PMMA), the *p*-isomer of methoxyphenamine and covered by the generic definition

Structure (6.49) shows 2-(4,7-dimethoxy-2,3-dihydro-1*H*-indan-5-yl) ethylamine, a substance not covered by the generic definition of a substituted phenethylamine, but listed by name as a Class A drug. It fails the generic test because the fused dihydroindan ring is not a univalent substituent.

Structure (6.49) 2-(4,7-Dimethoxy-2,3-dihydro-1*H*-indan-5-yl)ethylamine, a substance not covered by the generic definition of a substituted phenethylamine, but listed by name as a Class A drug

Fenfluramine (*N*-ethyl-α-methyl-3-trifluoromethylphenethylamine; Structure (6.50)), an anorectic drug, is also excluded from control; it has a haloalkyl ring-substitution that is neither halide nor alkyl.

Structure (6.50) Fenfluramine, an anorectic drug excluded by the generic definition and not listed specifically

A full list of 35 phenethylamines, which fell outside the generic definition, but were added to the Act as named substances in 2001, is shown in Appendices 17–19. Table 6.4 shows those ring-substituted

Table 6.4 Ring-substituted phenethylamines[a] reported in Europe since 1997

Substance name/PIHKAL Ref.	Name/Acronym
2-Chloro-4,5-methylenedioxymethylamphetamine (not listed in PIHKAL)[b]	Chloro-MDMA
4-Bromo-2,5-dimethoxy-N-ethylphenethylamine (not listed in PIHKAL)	N-Ethyl-2C-B
4-Bromo-2,5-dimethoxy-N-acetylamphetamine (not listed in PIHKAL)	N-Acetyl-DOB
1-(8-Bromobenzo[1,2-b;4,5-b'']difuran-4-yl)-2-aminopropane (not listed in PIHKAL)	Bromodragonfly
1-(8-Bromo-2,3,6,7-tetrahydrobenzo[1,2-b;4,5-b'']difuran-4-yl)-2-aminoethane (not listed in PIHKAL)	2C-B-Fly
4-Methoxy-N-ethylamphetamine (not listed in PIHKAL)	PMEA
4-Methylthioamphetamine (not listed in PIHKAL)	**4-MTA**
#7	ALEPH-7
#20	2C-B
#22	2C-C
#23	2C-D
#24	2C-E
#32	2C-H
#33	**2C-I**
#36	2C-P
#40	**2C-T-2**
#41	2C-T-4
#43	**2C-T-7**
#53	2,4-DMA
#64	DOC
#67	DOI
#107	MDHOET
#128	**MBDB**
#130	**PMMA**
#158	**TMA-2**

[a] Risk assessments were carried out on those substances shown emboldened (see Table 4.1)
[b] R.J. Lewis, D. Reed, A.G. Service and A.M. Langford, *The identification of 2-chloro-4, 5-methylenedioxymethylamphetamine in an illicit drug seizure*, J. For. Sci., 2000, **45(5)**, 1119–1125

phenethylamines that have been reported in Europe since 1997 that were not at the time under international control. Most were listed in PIHKAL. Apart from N-Acetyl-DOB, bromodragonfly and 2C-B-Fly, which are not controlled, all are Class A drugs. Of these, all are controlled by the generic definition of a substituted phenethylamine except MDHOET and 4-MTA, which are listed by name. Subsequently, 2C-B was added to Schedule I of UN1971.

6.14 TRYPTAMINES

Tryptamine (1H-indole-3-ethanamine) is a naturally occurring metabolite of the amino acid tryptophan, which, in turn, is a constituent of many proteins. Although tryptamine has few significant pharmacological properties, it forms the parent nucleus of a number of hallucinogenic drugs. Some of these are simple derivatives of tryptamine, whereas others are polycyclic structurally related substances such as β-carbolines and lysergamides. Many hallucinogenic tryptamines occur naturally in plants, fungi and, occasionally, animals, such as certain toads, but others are entirely synthetic or semi-synthetic substances.

A number of tryptamines are controlled by the Act as Class A drugs. Five are listed specifically in paragraph 1(a) of Part I of Schedule 2, namely bufotenine; etryptamine; psilocin; N,N-diethyltryptamine (DET) and N,N-dimethyltryptamine (DMT). Two structurally related substances containing the 2-(indol-3-yl)ethylamine fragment (corresponding to tryptamine) are also explicitly listed, namely lysergamide and lysergide. Other tryptamines are subsumed by the generic definition in paragraph 1(b) of Part I of Schedule 2, where Class A tryptamines are defined as: *"any compound (not being a compound for the time being specified in subparagraph (a) above) structurally derived from tryptamine or from a ring-hydroxy tryptamine by substitution at the nitrogen atom of the side-chain with one or more alkyl substituents but no other substituent"*. However, in Part I of Schedule 2, paragraph 3 provides for any ester or ether to be controlled. Taking both of these requirements together, then to qualify as a Class A drug, the following criteria must be satisfied (Structure (6.51)):

R^4, R^5, R^6 and R^7=H, OH, OR^x or O(CO)R^x where R^x includes alkyl, aryl, *etc.*
$R^1=R^2=R^{\alpha 1}=R^{\alpha 2}=R^{\beta 1}=R^{\beta 2}=H$
R'=H or alkyl
R''=alkyl

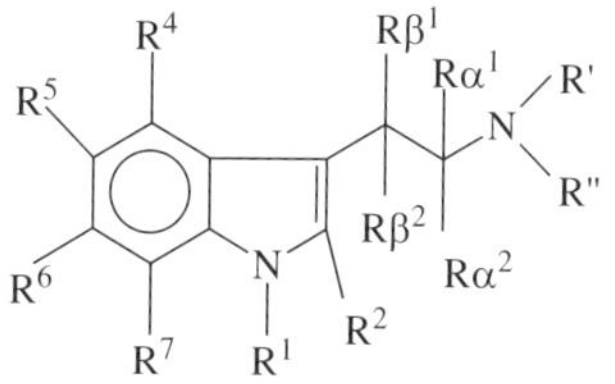

Structure (6.51) Tryptamine showing substitution patterns

The generic definition deliberately excludes from control tryptamines where there is no alkyl substitution on the side-chain nitrogen: a group that includes naturally occurring products such as serotonin and tryptamine itself.

6.14.1 The "TIHKAL" Drugs

A 1997 book following a similar format to the authors' previous publication (PIHKAL) is known by the acronym "TIHKAL" (see Bibliography); it provides synthetic monographs for 56 tryptamines together with notes on dosages, routes of administration, effects and properties of related compounds. Of these 56 substances, 9 are complex molecules containing the structure of tryptamine (*i.e.* four lysergamides, four β-carbolines and ibogaine). The remaining monographs include tryptamine itself and 45 derivatives. The monograph for lysergide (LSD; Substance #26 in TIHKAL) includes brief information on 43 other LSD derivatives, but these, the four substituted β-carbolines and ibogaine (a pentacyclic indole alkaloid) are not considered further here.

Table A20.1 shows the status of the TIHKAL tryptamine derivatives under the Act. Of the 56 substances featured in the monographs, then excluding tryptamine itself, the four β-carbolines and ibogaine (none of which is controlled in the UK or the UN Conventions)[9], there are 30 substances that are either listed specifically or defined generically as Class A drugs. In coming to this conclusion, it should be noted that a cyclobutyl substituent that leads to ring closure at the side-chain nitrogen atom (Compounds #24, #43 and #52) is not considered to be a N,N-dialkyl derivative; these three are therefore not controlled. On the other hand, a ring-dihydroxy group leading to a ring-diether (Substance #41) is regarded as satisfying the requirements.

Few of the tryptamines listed in TIHKAL have any actual or potential value to the pharmaceutical industry, but it should be noted that melatonin (Substance #35), although unlicensed in the UK, enjoys some status as a fringe medicine for the treatment of jet-lag and other sleeping disorders. Melatonin would not become a controlled drug by virtue of the extension of the generic definition shown below. There is yet little evidence that youth culture and the dance drug scene are likely to move away in the near future from the use of stimulants and empathogens typified by the phenethylamines towards hallucinogens of the tryptamine family. As well as not being stimulants, a more significant limitation is that many tryptamines are inactive when ingested. In order to produce an effect they must be smoked, injected or mixed with a monoamine oxidase inhibitor (MAOI). A good example of the latter situation is the hallucinogenic drink known as Ayahuasca or Caapi to certain indigenous people of South American. This is a concoction of plant extracts containing DMT (the hallucinogen) and harmine (the

---

[9] Ibogaine is a Schedule 1 substance in the US Controlled Substances Act

MAOI). Some tryptamines listed in TIHKAL appear to have no pharmacological action by any route.

Table A20.2 (Appendix 20) shows the 19 tryptamines that have been reported to EMCDDA since 1997 under the Early Warning System (Chapter 4), and that were not under international control. Seven of the 19 were not listed explicitly in TIHKAL and 4 are not controlled by the Act.

If, in the generic definition of a Class A tryptamine, the phrase *"ring-hydroxy tryptamine"* were to be replaced with *"ring-hydroxy or ring-alkylenedioxy tryptamine"*, then a further 5 drugs shown in Table A20.1 (Appendix 20; Substances #28 to #32) would be controlled. However, this would still leave other tryptamines uncontrolled. It might therefore be concluded that, because of the structural heterogeneity of the non-controlled tryptamines listed in TIHKAL or reported to the EMCDDA, specific nomenclature would be required to control them.

CHAPTER 7
Natural Products – Problem Areas

Nearly all controlled substances are chemically defined entities. Although a number of psychoactive drugs are found in a variety of plants, the only "vegetable" materials specifically controlled in the UK (Table 7.1) are coca leaf, opium, poppy-straw and its concentrate and fungi containing psilocin (all Class A) and cannabis and cannabis resin (both Class B from 2009). A separate offence of cultivation exists for these materials, although it is now common for cultivation to be subsumed under the broader offence of production of a controlled drug. Particular attention is given here to cannabis in its various forms since this has proved a troublesome area in forensic chemistry. This is followed by opium and poppy-straw, the identification and definition of the various forms of which continue to present problems, and finally "magic" mushrooms and coca tea.

7.1 CANNABIS AND CANNABIS RESIN

7.1.1 Introduction

In many ways, cannabis is central to drug control. It is the most widely consumed illegal substance in most countries. It has been used since antiquity, yet its pharmacology and harmful effects have only come into clear focus in recent years. This in turn has led to several recent shifts in its legal status. Overlaid on this has been the major problem of defining cannabis.

Forensic Chemistry of Substance Misuse: A Guide to Drug Control
By L.A. King
© L.A. King 2009
Published by the Royal Society of Chemistry, www.rsc.org

Table 7.1 Plants and plant products named in the Misuse of Drugs Act 1971 or the Drugs Act 2005.

Controlled drug	Botanical origin
Cannabis and cannabis resin	*Cannabis sativa*
Certain fungi containing psilocin	*Psilocybe* and other genera
Coca leaves	*Erythroxylon coca*
Poppy-straw, concentrate of poppy-straw and opium	*Papaver somniferum*

7.1.2 Definitions of Cannabis

Following the League of Nations Conventions of 1925 and 1931, control of cannabis (at that time commonly called Indian Hemp) had been restricted to the female plant. In the Dangerous Drugs Act, 1951, cannabis was defined as *"Indian hemp is the dried flowering and fruiting tops of the pistillate plant known as Cannabis sativa [alt. indica] from which the resin has not been extracted"*. The UN1961 Convention removed the exclusion of male plants and the restriction to a particular species of *Cannabis*, thereby avoiding taxonomic debate on whether or not that genus is monospecific. By 1971, the Misuse of Drugs Act defined cannabis (Section 37) as:

"[*"cannabis" (except in the expression "cannabis resin") means the flowering or fruiting tops of any plant of the genus Cannabis from which the resin has not been extracted, by whatever name they may be designated*]."

But problems soon arose and led to a number of contentious cases[1]. For example, the lower leaves are not part of the inflorescence, so presumably were not controlled, even though they may contain identifiable resin glands and THC. This caused difficulties in identifying finely divided material and smoking residues. A new definition was introduced by Section 52 of the Criminal Law Act (1977) as follows:

"[*"cannabis" (except in the expression "cannabis resin") means any plant of the genus Cannabis or any part of any such plant (by whatever name designated) except that it does not include cannabis*

---

[1] G.F. Phillips, *The legal description of cannabis and related substances*, Med. Sci. Law, 1973, **13(2)**, 139–142

resin or any of the following products after separation from the rest of the plant, namely –

(a) *mature stalk of any such plant,*
(b) *fibre produced from mature stalk of any such plant, and*
(c) *seed of any such plant;*]"

Even today, questions can still arise from the 1977 definition when herbal cannabis is examined. For example, what fraction of non-controlled stalk and seeds should a sample of cannabis contain before it is no longer considered to be controlled? Even if a sample comprising largely stalk and seeds is deemed to be controlled then, in calculating the weight, should some of the noncontrolled material be removed? This situation is reminiscent of the pre-1977 definition, described above, which excluded lower leaves.

7.1.3 Cannabis Seeds

Cannabis seeds are excluded from control, but despite "sterilisation", even seeds intended for use as bird food or fishing bait will often germinate. A new offence involving sale of seeds intended for cannabis cultivation and related cultivation equipment has been suggested, but has not so far been progressed. It is conceivable that cannabis seeds could be regarded as precursors to cannabis and therefore incorporated into the appropriate legislation. But apart from the uses noted above, cannabis seeds of approved type are produced on a large scale for the licensed cultivation of "low-THC" crops. These are intended for the manufacture of rope, paper and animal bedding; controls on seed would impact on this industry.

Although small amounts of cannabis (*e.g.* bracts and other flowering parts) may be found adhering to cannabis seeds, there are insignificant quantities of THC in clean seeds or in cannabis seed oil used for culinary or cosmetic purposes.

7.1.4 Hash oil

Cannabis (hash) oil has traditionally been made by solvent extraction of cannabis resin followed by removal of the excess solvent to leave a dark viscous liquid. It may contain ten times as much tetrahydrocannabinol (THC) as do cannabis and cannabis resin. Before 2003, the legal status of hash oil had developed into a notorious problem caused by the separate classification of vegetable matter (viz. cannabis and cannabis resin) in Class B with (pure) cannabinol derivatives in Class A. That distinction arose from the separate inclusion of the substances in the

UN1961 and UN1971 Conventions. The problem was solved immediately by the reclassification of cannabis, cannabinol and cannabinol derivatives into Class C (Class B from 2009). This saga, which is now of historical interest, is recounted in Appendix 6.

7.1.5 Cannabis-Based Medicines and Dronabinol

Nabilone, a synthetic analogue of THC, is licensed for hospital use in the UK as a treatment for the nausea arising from cancer chemotherapy, but not for other conditions (Structure (7.1)). It is not a controlled drug in the UK.

Structure (7.1) Nabilone

Naturally occurring tetrahydrocannabinol (THC) is the *R,R-trans*-stereoisomer of Δ^9-THC, *i.e.* {(6aR,10aR)-6a,7,8,10a-tetrahydro-6,6, 9-trimethyl-3-pentyl-6H-dibenzo[b,d]pyran-1-ol}. Dronabinol, the INN for THC, is not licensed for use in the UK, but may be imported on a "named patient basis", again for the same indications as nabilone; it is marketed as "Marinol" in the US. Dronabinol, including its stereoisomers, is listed in Schedule 2 of the Regulations. A decision by WHO in 2006 to move dronabinol from Schedule II to Schedule III of UN1971 was not supported by the UN Commission on Narcotic Drugs (CND).

Research is currently underway to evaluate the clinical potential of a number of cannabinoids extracted from cannabis plants. A proprietary preparation of cannabinoids, known as Sativex® and produced by GW Pharmaceuticals, is approved for use in Canada and is awaiting licensing approval in the UK from the Medicines and Healthcare products Regulatory Agency (MHRA). Originally it had been thought that Sativex® should be treated in the same way as dronabinol, *i.e.* placed in Schedule 2 of the Regulations. However, it can be argued that Sativex®, as an extract of cannabis, falls within the scope of the UN 1961 Convention under *"Extracts and Tinctures of Cannabis"*. It could therefore be listed in Schedule 4 of the Misuse of Drugs Regulations.

7.1.6 "High-Potency" Cannabis

The increasing THC content (potency) of certain types of cannabis was a major factor in both the 2005 and 2008 ACMD reviews of the classification of cannabis. From the point of view of a drug chemist, higher potency in itself causes no more difficulties than are already inherent in the analysis of cannabis, *i.e.* inhomogeneous samples, especially herbal cannabis, and the lack of accurate THC reference standards. Although it has been suggested from time to time that higher potency cannabis might have a different legal classification, this would be a sure route to legal problems. Not only is the precision (reproducibility) of quantitative cannabis assay worse than that for most other drugs of misuse, but such an approach could not target a particular type of cannabis since there is considerable overlap in the potencies of "traditional" (*i.e.* imported) herbal cannabis, cannabis resin and intensively cultivated herbal cannabis (sinsemilla/skunk)[2,3].

7.2 OPIUM

The different forms of opium are mentioned at various places in the Act:

- Paragraph 1 of Part I of Schedule 2 specifies *"Opium, whether raw, prepared or medicinal"* as being a Class A drug.
- Paragraph 5 of the same Part extends control to include *"any preparation or product containing [opium]"*.
- In Part IV of Schedule 2, there are two further definitions:
    (i) *"["medicinal opium" means raw opium which has undergone the process necessary to adapt it for medicinal use in accordance with the requirements of the British Pharmacopoeia, whether it is in the form of powder or is granulated or is in any other form, and whether it is or is not mixed with neutral substances;]"*
    (ii) *"["raw opium" includes powdered or granulated opium but does not include medicinal opium.]"*
- Section 37(1) offers a further definition:
"["prepared opium" means opium prepared for smoking and includes dross and any other residues remaining after opium has been smoked;]"

---

[2] L.A. King, C. Carpentier and P. Griffiths, *Cannabis potency in Europe*, Addiction, 2005, **100**, 884–886

[3] D.J. Potter, P. Clark and M.B. Brown, *Potency of Δ^9-THC and other cannabinoids in cannabis in England in 2005: Implications for psychoactivity and pharmacology*, J. For. Sci., 2008, **53(1)**, 90–94

The Regulations also refer to opium in several ways:

- Schedule 1 lists *"raw opium"*, and Schedule 2 lists *"medicinal opium"*, but prepared opium is not included.
- Paragraph 3 of Schedule 5 makes an exception from the prohibition on importation, exportation and possession (subject to the requirements of Regulations 24 and 25) for certain "low concentration" preparations of medicinal opium.
- A further exception is made in paragraph 8 of Schedule 5. This concerns mixtures of opium and ipecacuanha.

Finally, it should be noted that, apart from the general prohibitions on possession, possession with intent to supply, *etc.*, relating to all drugs in Schedule 2, there are specific offences within Sections 8 and 9 of the Act that refer to opium by name.

7.2.1 Definitions of Opium

The difficulty faced by the forensic scientist when dealing with suspected opium is that although raw, prepared and medicinal opium and opium preparations are apparently defined in the Act or Regulations, there is no clear statement of what constitutes opium. This problem is partly caused because opium is rarely seen in casework. As a consequence, analysts do not have the day-to-day familiarity that applies to, say, cannabis products. It is clear that confusion exists as to the distinguishing features of the different forms of opium. A second level of difficulty arises because definitions do exist, but which are unsuitable for the forensic chemist. For example, in The British Pharmacopoeia (BP), raw opium is defined as *". . . the air-dried latex obtained by incision from the unripe capsules of Papaver somniferum L. It contains not less than 10.0 per cent morphine . . . and not less than 2.0 per cent codeine . . . "*. The New Shorter Oxford English Dictionary defines opium as *"A reddish-brown strong-scented addictive drug prepared from the thickened dried juice of the unripe capsules of the opium poppy, used as a stimulant and intoxicant, and in medicine as a sedative and analgesic"*.

Recourse to the BP definition is unacceptable because it is not technically possible to identify the botanical origin of opium; the capsule exudates from certain other species of the *Papaver* genus also contain morphine and related alkaloids. Secondly, if a sample contains less than 10.0% morphine or less than 2.0% codeine then it is presumably not "opium of the Pharmacopoeia". It is therefore inappropriate that the legal definition should devolve onto a BP definition primarily developed to ensure the quality of an item of trade. The dictionary definition is similarly inadequate in that it refers to the origin and pharmacological properties of opium rather than its chemical constitution.

7.2.2 Identification of Opium

In the absence of clear criteria, forensic analysts opt to give an opinion as to whether a sample is opium-based on their knowledge and experience of the typical appearance and smell of opium. It is usually possible to demonstrate that the substance is a preparation or product containing a controlled drug by identifying the morphine present. However, this would not prove that the sample is opium, so opium-specific offences will not apply.

7.2.3 A New Definition of Opium?

A suitable definition of opium, which could be inserted into the Act, might be:

> *"Opium is a resinous material containing a range of alkaloids including morphine and codeine."*

However, even then it would not be entirely clear how an analyst would distinguish opium (made by capsule incision) from concentrate of poppy-straw (made by chemical extraction of poppy-straw), which is listed separately as a Class A drug in the Act.

7.3 POPPY-STRAW AND CONCENTRATE OF POPPY-STRAW

Part IV of Schedule 2 to the Act defines these as follows:

- "[*"poppy-straw"* means all parts, except the seeds, of the opium poppy, after mowing]."
- "[*"concentrate of poppy-straw"* means the material produced when poppy-straw has entered into a process for the concentration of its alkaloids]."

The dried seed head of an opium poppy, *i.e.* poppy-straw, often used in floristry, qualifies as a Class A drug, but by virtue of the Misuse of Drugs Regulations (Regulation 4), it is exempt from most controls relating to possession, production or supply.

There are no special exceptions for concentrate of poppy-straw. Not only is this material rarely seen, but it is not always clear analytically how it differs from some forms of opium.

7.4 "MAGIC" MUSHROOMS

The liberty cap mushroom *Psilocybe semilanceata* and certain other members of the *Psilocybe* and other fungal genera (*e.g. Amanita*) contain the hallucinogens psilocin (4-hydroxy-N,N-dimethyltryptamine) and its phosphate ester psilocybin, both of which are Class A drugs. Before 2005, the cultivation of so-called magic mushrooms had not been declared to be an offence either by statute or by case law. Whether a person is in possession of a controlled drug depends on whether they are in possession of a *preparation* containing a controlled drug. A general definition of "preparation" might be "any process that puts a substance or material into a form suitable for consumption". However, if the mushrooms had been treated before use or preserved in some way (*e.g.* deliberate drying, cooking or freezing) then those processes could be regarded as production of a Class A controlled drug. In R-v-Stevens, Cr.L.R. 568 (1981), the Court of Appeal rejected the notion that "preparation" had a technical pharmaceutical meaning. The court said: *"What was needed in order that these mushrooms should be prepared is that they ceased to be in their natural growing state and had in some way been altered by the hand of man to make them into a condition in which they could be used".* In R-v-Cunliffe, Cr.L.R. 547 (1986), the example was given of deliberate drying as an act of preparation. In the subsequent case of R-v-Hodder, Cr.L.R. 261 (1990), concerning frozen mushrooms, the possibility was opened up that such material could be considered an *"other product"* as defined in paragraph 5 of Part I of Schedule 2 to the Act.

Section 21 of the Drugs Act 2005 inserted into Part 1 of Schedule 2 to the Misuse of Drugs Act (*i.e.* Class A drugs), the following text: *"Fungus (of any kind) which contains psilocin or an ester of psilocin".* To a certain extent, this was a response to those legal problems of defining what constituted production, but Section 21 was also intended to counteract a rapid rise in the importation and sale of certain mushrooms, particularly *Psilocybe cubensis.* The Misuse of Drugs (Amendment) (No. 2) Regulations 2005 gave exemption from a possession offence when the mushrooms were growing naturally and were not being cultivated, or were being picked for the purposes of destruction. A European perspective on "magic mushrooms" was published in 2006[4].

As a result of Section 21 of the Drugs Act 2005, those earlier cases concerned with preparations of magic mushrooms are now to some extent only of historical interest. But they may continue to be relevant in dealing with the similar legal problems that can arise with peyote cactus, khat or other natural products that contain controlled drugs (Chapter 9). This has proved to be a difficult area for the law with continuing arguments over the definitions of terms such as preparation, production and product. The interested reader is directed to the more detailed treatment in *Misuse of Drugs: Offences, Confiscation and Money Laundering* – see Bibliography.

7.5 COCA TEA

Small bags of "coca de mate" (coca tea) are sometimes imported for personal use by travellers returning from South American countries. Although it is strictly coca leaf and therefore a Class A drug, there have been few if any prosecutions in cases of personal importation. The regular use of coca leaves is in any case almost unknown outside South America, and it could be argued that even in those countries the chewing of coca leaves by indigenous people is neither particularly harmful to individuals or the social structure. This is in stark contrast to the damage done to individuals and society by crack cocaine, and illustrates, in an extreme way, how the harmful properties of a substance depend on the physical and chemical form in which it is used.

---

[4] J. Hillebrand, D. Olszewski and R. Sedefov, *Hallucinogenic mushrooms: An emerging trend case study*, EMCDDA, 2006; http://www.emcdda.europa.eu/index.cfm?nNodeID = 7079

CHAPTER 8
Other Problems of Chemical/Legal Interpretation

8.1 CRACK COCAINE

Cocaine base is a white amorphous solid. When seen in the form of small lumps (rocks) it is known as crack, although this is a colloquial term without a clear scientific definition. Unlike cocaine hydrochloride, cocaine base can be smoked. Pure cocaine base can be crystallised as fine needles, but is never seen in this form. Cocaine, which includes crack, is a Class A controlled drug. In R-v-Russell (Appendix 8), it was held that the production of crack from cocaine (and by implication any salt/base interconversion) is an act of production for the purposes of Section 4 of the Act. In the absence of an allegation of production of crack, then a forensic analyst need only identify the constituent cocaine: a process that does not distinguish cocaine hydrochloride from crack. If this is then coupled with an estimate of the purity of the material, *i.e.* the percentage of cocaine base or base equivalent then no problem arises at the individual sample level. But if the frequency histogram of a number of such purity determinations of cocaine is then studied, it is usually found that the distribution curve is bimodal, *i.e.* a lower maximum (peak) closely relating to the modal purity of cocaine hydrochloride at around 40–50% and a higher maximum at around 70–80% caused by the presence in the sample population of cocaine base/crack. This arises because some samples of crack will be in powdered form and will have been described as cocaine.

A further aspect to be considered when examining crack is that its purity depends on the purity of the cocaine hydrochloride used in its

---

Forensic Chemistry of Substance Misuse: A Guide to Drug Control
By L.A. King
© L.A. King 2009
Published by the Royal Society of Chemistry, www.rsc.org

production and on the method of manufacture. Two processes may be used: In the first, alkali (*e.g.* sodium bicarbonate) is added to a hot aqueous solution of cocaine hydrochloride. This causes cocaine base to settle out as an oil that then solidifies and may be separated from the supernatant. When done carefully, this procedure can result in the purest crack. A simpler method is to mix solid cocaine hydrochloride with damp sodium bicarbonate and then heat in a microwave oven for a short period. In both methods, the amount of cocaine base is unaltered, but the microwave method leads to a lower-purity product simply because it is bulked up by sodium salts (*i.e.* chloride and residual carbonate/bicarbonate). Although rarely examined quantitatively, the presence and amount of specific adulterants can be used to profile crack. Such a profile could, in principle, be used to compare crack with the suspected parent cocaine as well as shedding light on the method of manufacture.

8.2 DIAGNOSTIC KITS

In the Misuse of Drugs Regulations (Regulation 2), certain products are exempted from controls. This includes so-called "diagnostic kits", which are commercial products used as calibrators in automatic drug-detection systems such as those for workplace drug testing. These may contain small amounts of controlled drugs. Such kits must first satisfy the requirements that they are not designed for administration of the drugs to a human or animal and that the drugs are not readily recoverable in a yield that constitutes a risk to health. They are then exempt provided that no one component part contains more than 1 mg of a controlled drug or 1 µg in the case of lysergide or any other *N*-alkyl derivative of lysergamide.

These threshold levels were designed to reduce the regulatory burden on manufacturers without opening an avenue for drug diversion. However, in framing the Regulation, some compromises were necessary; there are several substances where the required human dose is less than 1 mg, such that a kit containing these drugs could be misused. Although a number of rarely seen narcotic analgesics fall into this category, a less unusual substance is DOB (4-bromo-2,5-dimethoxy-α-methylphenethylamine), the effective dose of which is around 1 mg.

8.3 ISOTOPIC VARIANTS

The chemical elements (*e.g.* hydrogen, carbon, oxygen) each exist in a number of isotopic forms. These isotopes may be stable or unstable (radioactive). The isotopes of an element arise from the presence of different numbers of neutrons in their atomic nucleus. Their electronic structure and qualitative chemical properties are unchanged, but because they differ in mass, slight differences exist between the quantitative properties of the isotopes of a given element. Both stable and radioactive isotopes are widely used in analytical-chemical, diagnostic and other medical procedures.

The lightest element is hydrogen and it has three isotopes. The most abundant form (99.985%) is known simply as hydrogen or protium (^{1}H). The nucleus contains one proton. Deuterium (^{2}H or D) is also a stable isotope of hydrogen; it contains a proton and a neutron. The natural abundance of deuterium is 0.015%. A third isotope, known as tritium (^{3}H or T) contains two neutrons and is radioactive. Tritium has an extremely low abundance, and is normally manufactured in nuclear reactors. The enrichment of a chemical compound so as to increase the proportion of D is called deuteration. Ordinary carbon consists of two stable isotopes: 98.9% of mass twelve (^{12}C) and 1.1% of mass thirteen (^{13}C). Apart from hydrogen, isotopes of other elements do not have unique names or symbols.

8.3.1 A Case History

In a criminal trial in Sweden in the late 1990s, the defendants were charged with the unlawful manufacture of amphetamine. Their defence was that they intended to produce deuterated amphetamine, which was not a scheduled substance. After much debate and conflicting expert advice, the Supreme Court in Stockholm decided by a majority verdict on 5 July 1999 that deuterated amphetamine was to be considered as a substance under the Swedish Penal Law on Narcotics.

There is now an almost unanimous view that the specification of a particular isotope that forms part of a controlled substance does not influence the fact that the substance is subject to control. If practical matters are ignored, *i.e.* whether suitable precursors are available and if the process is economically attractive, then the main arguments in favour of this view are as follows:

- Isotopically pure substances do not exist. In the case of normal amphetamine, where there are nine carbon and thirteen hydrogen atoms, 0.2% of molecules will have at least one hydrogen replaced by deuterium and 10% of molecules will have at least one ^{13}C atom. An extreme example of this is afforded by the Class A controlled drug 4-bromo-2,5-dimethoxy-α-methylphenethylamine (DOB). Considering the bromine atom in this molecule and ignoring isotopic variation in

other atoms, then DOB exists in two almost equally abundant forms, *i.e.* a form containing ^{79}Br (50.6%) and a form containing ^{81}Br (49.4%). It can hardly be argued that one is controlled and the other not; the Act must cover both. If DOB were to be enriched or depleted in either bromine isotope, the product would still be controlled. By extension, the same argument applies to all isotopes.
- The biological properties of isotopic variants of controlled drugs differ only slightly from the normal compounds. There is no evidence that they cause any less social harm.
- Isotopic variants are not distinct chemical entities.

A dissenting view is based on a conservative interpretation of the UN Conventions, namely there is no explicit mention of these variants in the international drug control treaties.

8.3.2 Implications for the Misuse of Drugs Act

There remains a possibility that any future case involving isotopic variants could again lead to lengthy technical and legal arguments in a Court. If it were felt that, despite the above arguments, the status of isotopic variants should be clarified unambiguously in the Act, then new paragraphs could be added to Parts I, II and III of Schedule 2 referring to *"Any isotopic variants of a substance for the time being specified in paragraphs 1 or 2 (etc.) of this Part of this Schedule."*

8.4 LOW-DOSAGE PREPARATIONS

The purpose of Schedule 5 of the 2001 Regulations is to except a number of defined "low-dose" preparations from certain controls. The full text is set out in Appendix 3. Although designed to remove onerous restrictions on the medical and pharmaceutical professions, some aspects of Schedule 5 can prove troublesome for forensic scientists.

The first problem arises with the requirements of paragraph 1.(1), particularly in relation to dihydrocodeine. In the UK, proprietary preparations containing dihydrocodeine include DHC Continus®. This is available in 60 mg, 90 mg and 120 mg tablets. Because the dihydrocodeine is present as the bitartrate salt, even the 120 mg preparation contains only 80 mg of base; these tablets are therefore excepted by the provisions of paragraph 1.(1). But other dihydrocodeine tablets may be encountered where it is not necessarily obvious that the drug is present as the tartrate or any other specific salt. Since the origins of paragraph 1.(1) lie in UN1961, it is unlikely that manufacturers outside the UK would wish to market dihydrocodeine preparations containing more than 100 mg base, but if in doubt a quantitative analysis would have to be carried out.

The exemption of morphine and opium preparations containing less than 0.2% morphine base is not usually problematical. Apart from proprietary morphine–kaolin mixtures, which always have less than 0.2% morphine, the only other common preparations are morphine sulfate tablets. These always contain much more than 0.2% morphine. It should be noted that, in paragraph 3 of Schedule 5, the term *"preparation of ... morphine"* does not include derivatives of morphine. In R-v-Karagozlu (Inner London Crown Court, 2 December 1998), the defence argued successfully, but wrongly, that since the forensic analyst had not shown that the *diamorphine* concentration of a heroin exhibit was greater than 0.2%, there was no case to answer.

Apart from the exceptions listed in Schedule 5, there are no lower limits to the quantities of controlled drug required to create a possession offence. It is only necessary for the analyst to show unequivocally that the controlled drug is present. Legal arguments about what constitutes a usable amount of drug have largely been replaced by a need for the prosecution to show that the defendant was aware that he or she had possession of the drug.

An earlier exemption of cocaine preparations containing less than 0.1% cocaine was removed in 2005 (Chapter 5).

8.5 MEDICINAL PRODUCTS

The Class C anabolic steroids are subject to the provisions of Part II of Schedule 4 of the Regulations. They are excepted from the prohibition on possession as well as exportation and importation for personal use when in the form of a medicinal product. As noted in Chapter 2, the original definition of a medicinal product was set out in Section 130 (1) of the Medicines Act 1968, but this has been superseded by Article 1.2 of European Directive 2001/83/EC as amended by Directive 2004/27/EC. A medicinal product is:

(a) *"Any substance or combination of substances presented as having properties for treating or preventing disease in human beings; or"*.
(b) *"Any substance or combination of substances which may be used in or administered to human beings either with a view to restoring,*

correcting or modifying physiological functions by exerting a pharmacological, immunological or metabolic action, or to making a medical diagnosis".

The Misuse of Drugs Regulations 2001 note that "medicinal product" has the same meaning as in the Medicines Act 1968. In the absence of any subsequent amendment to the Regulations, that original definition must still apply. Many of the problems associated with controlled drugs in the form of medicinal products arose with the benzodiazepines. Schedule 4 of the Misuse of Drugs Regulations 1985 had excepted the benzodiazepines from the prohibition on possession when in the form of a medicinal product. However, this exception has been removed by the modified Schedule 4 of the Misuse of Drugs Regulations 2001. In cases where there might be a dispute between the Medicines Act and the Misuse of Drugs Act, the latter takes priority.

A number of uncertainties remain and the following examples have given rise to some debate:

- Illicitly-made tablets, capsules or injection ampoules would probably be recognised as medicinal products since the definition is not dependent on whether the manufacturer was or was not licensed to make the products.
- Although the pure chemical substance used as the raw material for the manufacture of tablets is not a medicinal product, the situation with crushed tablets may depend on the circumstances of an individual case.
- A view has been proposed that abusers of medicinal products containing anabolic steroids are not taking the drug for medicinal purposes, especially if the products are intended for veterinary use only, and therefore those steroids are no longer a medicinal product.

It should be noted that there is no requirement in the Regulations that to qualify as a medicinal product it must have been prescribed by a registered medical practitioner. Experience of the EU Early Warning System on "new psychoactive substances" (Chapter 4) has disclosed that Member States differ in their interpretation of European Directives. Thus, when illicit products containing the substituted piperazines *m*-chlorophenylpiperazine (*m*CPP) or 1-benzylpiperazine (BZP) (Chapter 9) became widespread, not all countries responded by invoking medicines legislation to restrict their manufacture and sale.

8.6 PURITY, POTENCY AND DRUG CONTENT

For basic drugs, the purity can be expressed as either the percentage of free base (anion) or the percentage of a particular salt. An exactly similar situation exists with acidic drugs (*e.g.* barbiturates) where the purity may be expressed in terms of free acid (cation) or a particular salt, but, as noted earlier, acidic drugs are uncommon. It is the general practice of forensic science laboratories to determine base drug purities. The reason for this is that unless a drug powder is known not to contain other ingredients, then it might be impossible to show that a specific salt is present. For example, suppose amphetamine is detected in a sample, but magnesium sulfate (Epsom salts) is also present as an intimate mixture. Detection of sulfate cations is unhelpful, but in the absence of further analysis, which is rarely justified, an analyst will not know if the amphetamine is in the form of amphetamine sulfate or some other salt of amphetamine. The analytical convenience of base purities is unfortunately marred by conceptual difficulties that can arise. Thus, under these circumstances, pure amphetamine sulfate has a (base) purity of only 73%, but a purity of 100% as amphetamine sulfate. The difference of 27% is accounted for by the sulfate part of the salt. The base purities of several drugs in the form of their pure salts are shown in Table A10.1.

Other difficulties can also arise in the statistical treatment of drug purities. Although the simple average, *i.e.* the arithmetic mean, is widely used and most easily understood, it is not always an ideal parameter for the central tendency of a population. In some circumstances other measures are necessary. One of these is the median, the value that divides the population into two such that there are as many values below the median as there are values above it. Another parameter is the mode, which describes the maximum, *i.e.* the most common or "typical" value for the distribution. Both the median and mode are appropriate measures when the population is skewed. An example here is the distribution of amphetamine purity; although most values cluster below 10%, the population has a long positive tail with a small number of high value purities. In this situation the common average will not only be much higher than the median or mode, but can be almost meaningless since few values will be close to the average. Finally, the weighted average is sometimes used when the population consists of samples of different purities and the weights of the individual samples vary widely. As an example, consider four samples of a powdered drug. Let the weights and purities be: 1 kg of 20%, 500 g of 18%, 5 g of 12% and 1 g of 10%. The mean purity (x) is simply the arithmetic average = {(20 + 18 + 12 + 10)/4} = 15%. The weighted average, (x_m) is the value that would be obtained if all four samples were thoroughly mixed and the purity remeasured. It is equal to the sum of the products of weight and purity divided by the total weight. For the four samples here, x_m is then $\{[(20 \times 1000) + (18 \times 500) + (12 \times 5) + (10 \times 1)]/1506\} = 19.3\%$. If $x_m > x$, then it suggests that sequential cutting is occurring as the drug is passed down a distribution chain and fragmented into smaller aliquots.

The concept of purity can only be applied to substances that are capable of existing in a pure state, even if that state is not often realised in practice. In other situations, drug potency is a more appropriate measure. The best example here is cannabis, where the active ingredient is THC. To say that a sample of cannabis has a purity of $x\%$, where $x\%$ is the THC content, is meaningless, because if the sample contained 100% THC it would no longer be cannabis. To put it another way, all cannabis is pure and to say otherwise suggests that it has been adulterated.

The term "drug content", while meaning either purity or potency in a fairly broad sense, is best used sparingly. It is ideally suited to the specific case of dosage units, particularly in a toxicological investigation. Thus, it is more helpful to say that a 250-mg tablet contains 75 mg of MDMA rather than that tablet has purity of 30%. This is because a tablet is usually consumed as a single entity, and the dose (*i.e.* 75 mg) is a useful measure of what has been ingested. Unless the weight of the tablet is stated, a purity of 30% conveys much less information. In practice, for both illicit and licensed tablets or other dosage units, the manufacturer will have endeavoured to make that unit with a fixed amount of drug, but the overall size and weight of that unit, and the presence of inert fillers, and hence the purity is irrelevant.

8.7 PREPARATIONS DESIGNED FOR ADMINISTRATION BY INJECTION

Paragraph 6 of Part I of Schedule 2 to the Act extends control to: *"Any preparation designed for administration by injection specified in any of paragraphs 1 to 3 of Part II of this Schedule."* In practice, amphetamine is the only commonly seen substance, which might qualify under this definition. In other words, does a syringe containing a solution of amphetamine cause that drug to be treated as a Class A rather than a Class B substance? There is no clear answer to this question; it appears that the issue is rarely prosecuted and has not been seriously tested in a trial. One view on this is that the above definition was inserted into the Act following abuse by injection of commercially produced methylamphetamine (*e.g.* Methedrine®) in the 1960s. It is less certain that paragraph 6 was meant to apply to *ad hoc* solutions of illicitly produced Class B drugs.

8.8 SUPPLY OF MEAT PRODUCTS

Anabolic/androgenic steroids are derivatives of testosterone, which is itself controlled. Thus, for the first time, the Act included a compound that occurs naturally in the tissues of both male and female humans, other mammals and birds. While possession of anabolic/androgenic steroids is not an offence, their unauthorised supply is illegal. This then raises a legal problem. If testosterone is present in human blood (the level is typically 5–10 micrograms/L), then providing transfusion blood could be deemed to be supply. In reality this issue is not new, since who can be sure that transfusion blood never contains any other controlled substance? This is in any case a somewhat academic point since a prosecution is not likely to be in the public interest. A similar question arises with the supply of meat products; they also contain testosterone. The levels may be low, but the Act makes no current provision for a *de minimis* approach to steroids.

CHAPTER 9
Candidates for Future Control

9.1 NEW PSYCHOACTIVE SUBSTANCES IN THE "POST-PIHKAL" ERA

It is difficult to overestimate the importance of the book PIHKAL in generating interest in synthetic drugs. The years after its publication in 1991 saw a huge rise in the production of illicit phenethylamines: substances that are covered in Chapter 6. The impact of a sister publication TIHKAL was less marked; although many illicit products containing novel tryptamines have appeared since 1997, they have been of minor significance. The "phenethylamine" period lasted until well into the 21st century, but now appears to be largely exhausted. The past few years have seen few new phenethylamines, but a huge diversification into new drug families. Two major series have already appeared; the substituted piperazines and the substituted cathinones. But substitution in even apparently minor compounds such as aminoindan could lead to new series of illicit drugs. This is a continuation of an historical process, whereby restrictions on a substance or its precursor chemicals stimulate clandestine chemists to explore new compounds, which in turn lead to further controls. And that cyclic process seems to be gathering momentum. Such developments will almost certainly generate new legislative and analytical challenges.

Generic controls have proved useful in the past for anticipating misuse of novel substances, but the current diversity of compounds puts severe strains on this approach. Whilst it may seem difficult to anticipate the next substance to appear on the illicit market, some predictions can be made.

Forensic Chemistry of Substance Misuse: A Guide to Drug Control
By L.A. King
© L.A. King 2009
Published by the Royal Society of Chemistry, www.rsc.org

Despite the various chemical groups involved, certain trends can be identified. Thus, not only is there a continuing effort to find novel, non-scheduled CNS stimulants, but the structures involved so far continue to explore the theme of an aromatic ring (usually phenyl) bearing a side-chain with an amino group, which may be primary, secondary or tertiary. The amino nitrogen is separated from the ring by one or, more usually, two carbon atoms, and the side-chain may be free or cyclic (*i.e.* part of a second ring). This structural pattern is not unique to stimulants, but it was originally exploited by chemists when creating the phenethylamine and tryptamine family. It can be seen in 1-benzylpiperazine, the "cathinones", and in derivatives of indan, indene and tetralin. It is also present in what may be cautiously termed "obsolete stimulants" such as aminorex (2-amino-5-phenyl-2-oxazoline; Structure (9.1)), 4-methylaminorex and pemoline (2-imino-5-phenyloxazolidin-4-one; Structure (9.2)).

Structure (9.1) Aminorex

Structure (9.2) Pemoline

As with the phenethylamines, ring-substitution into a structural prototype may lead to substances that mimic the effects of MDMA. Although two new phenethylamines have recently appeared, neither bromodragonfly nor 2C-B-Fly (see below) was described in PIHKAL. The latter two compounds are hallucinogenic, but it is noteworthy that they and other synthetic hallucinogens remain on the fringes.

Apart from variations on the arylalkylamine nucleus, this chapter covers a miscellaneous group of substances, where it is difficult to ascertain any single theme. The substances listed in the following sections are mostly those that have come to notice in drugs seizures by law-enforcement agencies in Europe or have been sold via internet websites as "legal highs". Some are licensed medicinal products that have been misused or have appeared in unlicensed formulations. It appears that some pharmaceutical agents, when administered by a different route or in amounts larger than required for normal therapy, may give rise to unexpected psychoactive effects. This group includes, for example, dextromethorphan, glaucine and benzydamine. It is difficult to predict which of these agents are likely to become established drugs of misuse and those which represent a short-lived experimentation. It could be argued that some of the substances described in the following sections are now only of historical interest. But insofar as they have been misused and are not controlled, they could reappear at any time. As with many "new" substances, their pharmacology is often unclear.

9.2 OTHER PHENYLALKYLAMINES

The phenethylamine nucleus has been a particularly fruitful source of new synthetic substances. Apart from the ring-substituted members (*i.e.* MDMA, *etc.*), variations in the side-chain have given rise to three series: (a) those that are *N*-substituted, (b) phenylalkylamines other than 2-phenylethylamines, and (c) more complex phenethylamines. Examples from all three groups have occurred in drug seizures, and are described below.

9.2.1 *N*-Substituted Phenethylamines

The *N*-substituted phenethylamines make up a rather mixed group. Some are well known as established drugs of abuse while others have value in medicine with little abuse potential. Many of the latter are still of forensic interest because they metabolise to either amphetamine or methylamphetamine and may be detected in the urine[1]. Nearly all of these compounds are *N*-substituted α-methyl-phenethylamines (*i.e. N*-substituted amphetamines). One of the simplest is methylamphetamine (Class A): the *N*-methyl derivative of amphetamine. Structure (9.3) shows the general form of a *N*-substituted amphetamine. Table 9.1 lists some of the better-known substances and a few examples of compounds that have been found in seizures. The Class C drug mesocarb is excluded from Table 9.1 because it is not a simple *N*-substituted phenethylamine; the amine nitrogen is part of a ring structure. The attraction of certain *N*-substituents to an illicit chemist is that a noncontrolled drug can be made that is sufficiently labile that it can be converted metabolically or by other simple means into an active substance: in other words, it is a proxy for a controlled drug. A good example here is

---

[1] J.T. Cody, *Metabolic precursors to amphetamine and methamphetamine*, For. Sci. Rev., 1993, **5(2)**, 109–127

Table 9.1 N-substituted amphetamines – see Structure (9.3).

Compound	R^1	R^2	Comments
N-Acetylamphetamine	H	Acetyl	Illicit product
Amphetamine	H	H	Class B controlled drug
Amphetaminil	H	1-Phenyl-1-cyanomethyl	Medicinal product (not UK)
Benzphetamine	H	Benzyl	Class C controlled drug
Clobenzorex	H	o-Chlorobenzyl	Medicinal product (not UK)
N,N-Dimethylamphetamine	Methyl	Methyl	Illicit product
N,N-Di-(2-phenylisopropyl)amine	H	2-phenyl-isopropyl	Illicit product known as DPIA
N-Ethylamphetamine	H	Ethyl	Class C controlled drug
Famprofazone	Methyl	3-(1-Phenyl-2-methyl-4-isopropylpyrazolin-5-one)methyl	Medicinal product (not UK)
Fencamine	Methyl	See footnote[a]	Medicinal product (not UK)
Fenethylline	H	7-Theophyllinylethyl[b]	Class C controlled drug
Fenproporex	H	2-Cyanoethyl	Class C controlled drug
Furfenorex	Methyl	2-Furylmethyl	Medicinal product (not UK)
N-(2-Hydroxyethyl)amphetamine	H	2-Hydroxyethyl	Illicit product
Mefenorex	H	3-Chloropropyl	Class C controlled drug
Methylamphetamine	H	Methyl	Class A controlled drug
α-Methylphenethyl-hydroxylamine	H	Hydroxy	Class B controlled drug
Prenylamine	H	3,3-Diphenylpropyl	Medicinal product (not UK)
Selegiline	Methyl	2-Propynyl	Medicinal product (UK)

[a]The R^2 substituent in fencamine is 3,7-dihydro-1,3,7-trimethyl-1H-purine-2,6-dione-8-amino-2-ethyl

[b]The term "theophyllinyl" refers to 3,7-dihydro-1,3-dimethyl-1H-purine-2,6-dione

α-methylphenethylhydroxylamine (the N-hydroxy derivative of amphetamine), now listed as a Class B drug.

Structure (9.3) The general structure of a N-substituted amphetamine

9.2.2 Other Side-Chain Phenylalkylamine Variants

In phenethylamine, the amino group is separated from the phenyl ring by two saturated carbon atoms. This configuration appears to be

Table 9.2 Derivatives of 1-phenylethylamine – see Structure (9.4).

Name	R^1	R^2
N-Methyl-1-phenylethylamine	Methyl	H
4-Methyl-1-phenylethylamine	H	Methyl
1-Phenylethylamine (α-methylbenzylamine)	H	H

optimal for pharmacological activity. However, illicit chemists have experimented with other arrangements. In the mid-1990s, 1-phenylethylamine (α-methylbenzylamine; Structure (9.4)) and, less commonly, the isomeric 4-methyl and N-methyl analogues appeared in drug seizures in Europe. In the same period, a further ring-substituted compound (3,4-methylenedioxy-N-methylbenzylamine) was found in the Netherlands and Germany. Although not described in detail, the latter is mentioned in PIHKAL under the code name ALPHA. As far as is known, these compounds (Table 9.2) behave as weak stimulants, but the pharmacology of the ring-substituted derivatives (*e.g.* ALPHA) might be closer to MDMA. A single example (1-phenyl-3-butanamine, also known as homoamphetamine; Structure (9.5)) has been encountered where the amino group is more distant from the phenyl group. None of these compounds is controlled either in the UK or in the UN Conventions. Two related compounds, N-benzylmethylamine and N-benzylethylamine have been reported in the US as mimics for crystalline methylamphetamine. Neither is controlled and neither shows any noticeable effects on the CNS[2].

Structure (9.4) 1-Phenylethylamine showing substitution patterns

Structure (9.5) 1-Phenyl-3-butanamine

---

[2] Anon. *N-benzylmethylamine HCl and N-benzylethylamine HCl ("Ice" and crystal methamphetamine mimics) in the Southwest*, DEA Microgram Bulletin, 2007, **40(8)**, 79–80

9.2.3 Other Ring-Substituted Phenethylamines

Two phenethylamines, known as bromodragonfly {1-(8-bromobenzo[1,2-*b*;4,5-*b"*]difuran-4-yl)-2-aminopropane} and the analogous 2C-B-Fly {1-(8-bromo-2,3,6,7-tetrahydrobenzo[1,2-*b*;4,5-*b"*]difuran-4-yl)-2-aminoethane}, have appeared in drug seizures in several European countries. Their trivial names reflect the pictorial representation of the fused rings shown in Structures (9.6) and (9.7). The two substances are not homologues since the furanyl rings in 2C-B-Fly are saturated. Neither is listed in PIHKAL and neither is scheduled under UN1971. They are not covered by the generic definition of a substituted phenethylamine (Chapter 6) since the fused furanyl rings do not constitute permitted substituents. However, bromodragonfly is controlled in Denmark and Sweden. Both substances are potent and long-lasting hallucinogens, somewhat similar to lysergide (LSD) with doses in the submilligram range. It has been suggested that the *R*-stereoisomer is the most active in both cases. Analytical properties for 2C-B-Fly, bromodragonfly and 3C-B-Fly (the α-methyl derivative of 2C-B-Fly) have been recently described[3].

Structure (9.6) Bromodragonfly

Structure (9.7) 2C-B-Fly

9.3 1-BENZYLPIPERAZINE AND OTHER DERIVATIVES OF PIPERAZINE

1-benzylpiperazine (BZP) is a CNS stimulant with about 10% of the potency of d-amphetamine. Together with other piperazine derivatives, BZP

---
[3] E.C. Reed and G.S. Kiddon, *The characterization of three FLY compounds (2C-B-FLY, 3C-B-FLY and Bromo-dragonFLY)*, Microgram Journal, 2007, **5(1–4)**

had become a popular substance of misuse in New Zealand. It is mentioned briefly in PIHKAL (see Bibliography). Although first reported in Europe in 1999, it became much more prevalent after 2004. The EMCDDA carried out a risk assessment on BZP in mid-2007 and concluded that it should be controlled within the EU[4]. In early 2008, the European Council issued a formal notification[5] to Member States that BZP should be made subject to control measures and criminal provisions in accordance with their national law, as provided for under their legislation.

But BZP is only one of several substituted piperazines, misuse of which has been reported in the EU and elsewhere in recent years. Thus, 1-(3-chlorophenyl) piperazine (*m*CPP), has been even more widespread than BZP. By 2006, it was estimated that almost 10% of illicit tablets sold in the EU, as part of the illicit ecstasy market, contained *m*CPP. Apart from *m*CPP, the next most commonly found substituted piperazine was 1-(3-trifluoromethyl-phenyl)piperazine (TFMPP), although it was nearly always seen in combination with BZP.

9.3.1 *N*-Substituted Piperazines: A Possible Generic Definition

There are at least 12 substituted piperazines that have been reported in Europe or the US in recent years. They can be divided into the 1-phenyl series and the 1-benzyl series as shown in Structure (9.8) and Table 9.3, and Structure (9.9) and Table 9.4, respectively. Whereas BZP is a stimulant, the other substituted piperazines have a more complex pharmacology; some appear to mimic or potentiate the effects of MDMA. Although the UK is only required to implement controls on BZP, it is suggested that a group of substituted piperazines could be brought under generic control. A possible definition might read as follows:

> "*N-benzylpiperazine and any compound structurally derived from N-benzylpiperazine or N-phenylpiperazine by substitution in the aromatic ring to any extent with alkyl, alkoxy, alkylenedioxy, halide or haloalkyl substituents, whether or not substituted at the second nitrogen atom of the piperazine ring with alkyl, benzyl, haloalkyl, or phenyl substituents.*"

The above definition captures the substances listed in Tables 9.3 and 9.4. However, it is essential that the definition does not inadvertently

---

[4] EMCDDA, *Report on the risk assessment of BZP in the framework of the Council Decision on new psychoactive substances* (in press)
[5] Council Decision 2008/206/JHA; http://www.emcdda.europa.eu/html.cfm/index875EN.html

Table 9.3 Phenylpiperazines – see Structure (9.8).

Name (Acronym)	R^1	R^2	R^3	R^4
1-(3-Chlorophenyl)-4-(3-chloropropyl)piperazine (mCPCPP)	H	Cl	H	$CH_2CH_2\text{-}CH_2Cl$
1-(3-Chlorophenyl)piperazine (mCPP)	H	Cl	H	H
1-(4-Chlorophenyl)piperazine (pCPP)	Cl	H	H	H
1-(4-Fluorophenyl)piperazine (pFPP)	F	H	H	H
1-(2-Methoxyphenyl)piperazine (oMeOPP)	H	H	MeO	H
1-(4-Methoxyphenyl)piperazine (pMeOPP)	MeO	H	H	H
1-(3-Methylphenyl)piperazine (mMPP)	H	Methyl	H	H
1-(4-Methylphenyl)piperazine (pMPP)	Methyl	H	H	H
1-(3-Trifluoromethylphenyl)piperazine (TFMPP)	H	CF_3	H	H

Table 9.4 Benzylpiperazines – see Structure (9.9).

Name (Acronym)	R^4
1-Benzyl-4-methylpiperazine (MBZP)	Methyl
1-Benzylpiperazine (BZP)	H
1,4-Dibenzylpiperazine (DBZP)	$C_6H_5\text{-}CH_2$

subsume an active pharmaceutical ingredient (API), a number of which are based on substituted piperazine. The most closely related group of APIs derived from piperazine include cyclizine (1-diphenylmethyl-4-methylpiperazine) and its many derivatives. None of these, nor more distantly related substances (*e.g.* diethylcarbamazine, vanoxerine, and trazodone) falls within the above definition.

Two substances listed in Table 9.3 do have legitimate applications. Thus, *m*CPP is used as a probe of serotonin receptors in experimental neuropharmacology and as the precursor in the synthesis of the anti-depressant drug trazodone. The substance 1-(3-chlorophenyl)-4-(3-chloropropyl)-piperazine (*m*CPCPP) is a precursor used in the manufacture of the anti-depressant drug nefazodone. If these substances were to become controlled drugs by virtue of the above generic definition then two options exist: either a licence could be issued to those using them for scientific/industrial purposes or specific exclusions could be made in the definitions. However, it is not certain that either *m*CPP or *m*CPCPP is used commercially in the UK. Little is known about the pharmacology of the three di-*N*-substituted compounds (1-(3-chlorophenyl)-4-(3-chloropropyl)piperazine,1,4-dibenzylpiperazine and 1-methyl-4-benzylpiperazine). They may have limited potential for misuse since the absence of a secondary amino group might reduce or eliminate pharmacological activity. DBZP is a known synthetic impurity in the manufacture of BZP, and is often found in association with BZP in illicit products, while there have only been two reports in the EU of the occurrence of 1-methyl-4-benzylpiperazine.

In 2008, ACMD proposed that BZP and generically defined piperazines should become Class C drugs, in line with the limited evidence of their harmfulness. They should be listed in Schedule 1 of the Regulations in view of their lack of medicinal value. In the EU, eight countries control *m*CPP (Belgium, Cyprus, Denmark, Germany, Hungary, Lithuania, Malta and Slovakia), and control is pending in Bulgaria. As noted in Chapter 10, BZP and five other piperazines are now controlled in New Zealand.

Structure (9.8) The general structure of a substituted phenylpiperazine

Structure (9.9) The general structure of a substituted benzylpiperazine

9.4 SUBSTITUTED CATHINONES

Cathinone (a Class C drug) could form the basis of an equally large series of novel compounds. Cathinone occurs as a natural constituent of khat (Chapter 2), and can be described as the β-keto analogue of amphetamine. The *N*-methyl homologue (methcathinone; Class B) and dimethylcathinone are wholly synthetic; they are the β-keto analogues of methylamphetamine and dimethylamphetamine, respectively. The *N,N*-diethyl derivative of cathinone is diethylpropion (amfepramone; Class C): a substance once used widely as an anorectic, but also abused for its stimulant properties. Bupropion, the 3-chloro-*N*-t-butyl derivative of cathinone, is an anti-depressant drug used in the treatment of smoking dependence. Methylone (3,4-methylenedioxymethcathinone; MDMCAT), the β-keto analogue of MDMA, has been widely seen. Ethylone is the corresponding analogue of MDEA, and βk-MBDB is the analogue of MBDB. Several derivatives, where the amino group of cathinone has been absorbed into a pyrrolidine ring, are closely related to pyrovalerone, a Class C controlled drug, listed in Schedule IV of UN1971. One of them (PPP) is closely related to prolintane (1-(1-phenylpentan-2-yl)pyrrolidine), a substance that was removed from the Misuse of Drugs Act in 1973. The clandestine manufacture of ring-substituted cathinones had been anticipated by the US Drug Enforcement Administration as early as 1997[6]. The illicit cathinone derivatives with α-pyrrolidino-substitution appeared in Germany in recent years and have been extensively described[7,8]. Both mephedrone (4-methylmethcathinone; 2-methylamino-1-p-tolylpropan-1-one) and ethcathinone are homologues of methcathinone; they had been offered for sale on the Internet[9] and were subsequently discovered in capsules seized by Finnish customs in early 2008. Described as "Sub-Coca", they had allegedly been sold by a company in Israel, although Internet references[10] make no mention of the active ingredients. The synthesis and properties of *N*,*N*-dimethylcathinone and *N*-ethylcathinone were recently described[11].

All of these complex derivatives of cathinone probably have a similar pharmacological activity to their corresponding phenethylamine analogues. A study of a series of analogues of pyrovalerone (*i.e.* pyrrolidinylphenyl-pentanones) showed that they were selective inhibitors of dopamine and noradrenaline transporters[12].

---

[6] T.A. Dal Cason, *The characterization of some 3,4-methylenedioxycathinone (MDCATH) homologs*, For. Sci. Int., 1997, **87**, 9–53
[7] F. Westphal, T. Junge, P. Rösner, G. Fritschi, B. Klein and U. Girreser, *Mass spectral and NMR spectral data of two new designer drugs with an α-aminophenone structure: 4'-Methyl-α-pyrrolidinohexanophenone and 4'-methyl-α-pyrrolidinobutyrophenone*, For. Sci. Int., 2006, **169(1)**, 32–42
[8] D. Springer, F.T. Peters, G. Fritschi and H.H. Maurer, *New designer drug 4'-methyl-α-pyrrolidinohexanophenone: studies on its metabolism and toxicological detection in urine using gas-chromatography–mass spectrometry* J. Chromatog. B, 2003, **789(1)**, 79–91
[9] http://www.ogc.dk/. See also http://www.erowid.org/chemicals/ethylcathinone/ethylcathinone.shtml and http://www.erowid.org/chemicals/4_methylmethcathinone/4_methylmethcathinone.shtml
[10] http://wikihighs.com/index.php?title = Sub_Coca and http://www.feedmybush.com/index.php?searchStr = sub+coca&act = viewCat&submitHidden = Go
[11] T.A. Dal Cason, *Synthesis and identification of N,N-dimethylcathinone hydrochloride*, Microgram Journal, 2007, **5(1–4)**
[12] P.C. Meltzer, D. Butler, J.R. Deschamps and B.K. Madras, *1-(4-Methylphenyl)-2-pyrrolidin-1-yl-pentan-1-one (Pyrovalerone) Analogues: A promising class of monoamine uptake inhibitors*, J. Med. Chem., 2006, **49**, 1420–1432

Cathinone derivatives without 3,4-methylenedioxy ring-substitution are shown in Table 9.5 (see also Structure (9.10)), while cathinone derivatives with 3,4-methylenedioxy ring-substitution are shown in Table 9.6 (see also Structure (9.11)). None of the 3,4-methylenedioxy cathinones shown in Table 9.6 are under UK or international control. Where appropriate, the corresponding phenethylamine analogue is also listed. Many of the substances shown in Tables 9.5 and 9.6 have appeared as illicit products in the EU or have been offered for sale via European websites. Little is known about whether the illicit cathinones are racemic mixtures or have been produced by stereoselective synthesis; information on the relative potency of the enantiomers is also lacking.

Table 9.5 Cathinone and various derivatives – see Structure (9.10).

R^1	R^2	R^3	R^4	R^5	Name (Acronym) [phenethylamine analogue]	Control status UK/UN Schedule
H	H	H	H	H	Cathinone [*amphetamine*]	Class C UN1971(I)
Methyl	H	H	H	H	Methcathinone [*methylamphetamine*]	Class B UN1971(I)
Methyl	Methyl	H	H	H	*N,N*-Dimethylcathinone [*dimethylamphetamine*]	Not controlled
Ethyl	H	H	H	H	*N*-Ethylcathinone [*ethylamphetamine*]	Not controlled
Methyl	H	4-Methyl	H	H	Mephedrone [*4,N-dimethylamphetamine*]	Not controlled
Ethyl	Ethyl	H	H	H	Diethylpropion (amfepramone) [*N,N-diethylamphetamine*]	Class C UN1971(IV)
t-Butyl	H	3-Cl	H	H	Bupropion [*3-chloro-N-t-butylamphetamine*]	Not controlled – used in a medicinal product in UK
{pyrrolidino}		H	H	H	α-Pyrrolidinopropiophenone (PPP)	Not controlled
{pyrrolidino}		4-Methyl	H	H	4-Methyl-α-pyrrolidinopropiophenone (MPPP)	Not controlled
{pyrrolidino}		4-MeO	H	H	4-Methoxy-α-pyrrolidinopropiophenone (MOPPP)	Not controlled
{pyrrolidino}		4-Methyl	Propyl	H	4-Methyl-α-pyrrolidinohexanophenone (MPHP)	Not controlled
{pyrrolidino}		4-Methyl	Ethyl	H	Pyrovalerone; 4-methyl-α-pyrrolidinovalerophenone	Class C UN1971(IV)
{pyrrolidino}		4-Methyl	Methyl	H	4-Methyl-α-pyrrolidinobutyrophenone (MPBP)	Not controlled
{pyrrolidino}		4-Methyl	H	Methyl	4-Methyl-α-pyrrolidino-α-methylpropiophenone	Not controlled

Table 9.6 3,4-Methylenedioxy-derivatives of cathinone – see Structure (9.11).

R^1	R^2	R^4	Name (Acronym or common name); [phenethylamine analogue]
Methyl	H	H	3,4-methylenedioxyphenyl-2-methylamino-1-propanone (Methylone; βk-MDMA); [$MDMA$]
Ethyl	H	H	3,4-methylenedioxyphenyl-2-ethylamino-1-propanone (Ethylone; βk-MDEA); [$MDEA$]
Methyl	H	Methyl	3,4-methylenedioxyphenyl-2-methylamino-1-butanone (Butylone; βk-MBDB); [$MBDB$]
{pyrrolidino}		H	3,4-Methylenedioxy-α-pyrrolidinopropiophenone (MDPPP)
{pyrrolidino}		Ethyl	3,4-Methylenedioxypyrovalerone (MDPV)

Table 9.5 shows the legal status of cathinone derivatives with respect to UN1971 and the Misuse of Drugs Act. In addition, methylone is controlled in Denmark and Sweden.

Structure (9.10) Substitution patterns in cathinone derivatives

Structure (9.11) Substitution patterns in 3,4-methylenedioxy-derivatives of cathinone

9.5 SUBSTITUTED INDANS, INDENES AND TETRALINS

In Europe, 2-aminoindan (Structure (9.12)) has been reported in drug seizures. It is a short-acting stimulant with effects that have been compared to 1-benzylpiperazine or methylamphetamine. From a structural aspect, 2-aminoindane is closely related to amphetamine, where the α-methyl substituent has been connected to the aromatic ring. In 1999, a chemical company in the UK received enquiries from Sweden regarding synthesis of the related compounds 5,6-methylenedioxy-2-aminoindene (Structure (9.13)) and 5,6-methylenedioxy-2-aminotetralin (Structure (9.14)). Neither was seen in circulation and little is known of their pharmacological properties. Although the former would show unsaturation in the side chain, these compounds are essentially the α-alkyl ring-closed analogues of MDA and BDB (1-[1,3-benzodioxol-5-yl]-2-butanamine; PIHKAL ≠94). The synthesis and pharmacological properties of the benzofuran, indan and tetralin analogues of MDMA have been described[13]. A chemical company with a European website[14] has indicated that they are considering the sale of 1-(5-indanyl)-2-aminopropane (Structure (9.15)). Otherwise known as the indanyl analogue of MDA (5-IAP), it has already been submitted to forensic science laboratories in the US as suspected ecstasy[15]. On a structural basis, it is possible that 5-IAP acts as a CNS stimulant. Other conformationally restricted phenethylamine analogues have been evaluated as 5-HT receptor agonists where the aminoalkyl side-chain has been partly converted to a cyclic structure. It was predicted that the R-enantiomer of the benzocyclobutene analogue of 2C-B would be the most potent[16]. These substances are all potential candidates for illicit production.

Structure (9.12) 2-Aminoindan

Structure (9.13) 5,6-methylenedioxy-2-aminoindene

Structure (9.14) 5,6-methylenedioxy-2-aminotetralin

Structure (9.15) 1-(5-Indanyl)-2-aminopropane

---

[13] A.P. Monte, D. Marona-Lewicka, N.V. Cozzi and D.E. Nichols, *Synthesis and pharmacological examination of benzofuran, indan and tetralin analogues of 3,4-(methylenedioxy)amphetamine*, J. Med. Chem., 1993, **36**, 3700–3706
[14] http://www.ogc.dk/page008.html
[15] J.F. Casale, T.D. McKibben, J.S. Bozenko and P.A. Hays, *Characterization of the "indanylamphetamines"*, Microgram Journal, 2005, **3(1–2)**
[16] T.H. McLean, J.C. Parrish, M.R. Braden, D. Marona-Lewicka, A. Gallardo-Godoy, and D.E. Nichols, *1-Aminomethylbenzocycloalkanes: conformationally restricted hallucinogenic phenethylamine analogues as functionally selective 5-HT$_{2A}$ receptor agonists*, J. Med. Chem., 2006, **49**, 5794–5803

9.6 γ-BUTYROLACTONE (GBL), 1,4-BUTANE-DIOL (1,4-BD) AND RELATED SUBSTANCES

As discussed in Chapter 5, the substance 4-hydroxy-n-butyric acid (GHB) exists not only in a salt/acid equilibrium – both of which forms are controlled – but the free acid is also in equilibrium with the lactone (γ-butyrolactone: GBL). When ingested, GBL forms GHB, yet GBL is not currently controlled. A Europe-wide review of the misuse of GHB and GBL was recently published[17]. Closely related to GBL, 1,4-butane-diol (1,4-BD; Structure (9.16)) is also metabolised to GHB. Although they are both widely used as industrial solvents, plans are in hand for GBL and 1,4-BD to join GHB as Class C drugs. Although γ-valerolactone (5-methyldihydrofuran-2(3H)-one; GVL), metabolises to γ-hydroxyvalerate (methyl-GHB; GHV), which has similar physiological effects to GHB, it has remained uncommon.

$$HO-CH_2-CH_2-CH_2-CH_2-OH$$

Structure (9.16) 1,4-butane-diol (1,4-BD)

9.7 EPHEDRINE AND PSEUDOEPHEDRINE

Ephedrine, either as a synthetic substance or in the form of extracts of *Ephedra vulgaris* (known as Ma Huang in Chinese medicine), is used and abused in several different ways. As a medicine it finds wide application as a bronchodilator to treat bronchospasm associated with asthma, bronchitis and emphysema. It is abused for its stimulant properties, but l-ephedrine is 5 times less potent than amphetamine, although somewhat more potent than diethylpropion (amfepramone). Pseudoephedrine is used as a decongestant. Both are precursors in the clandestine synthesis of methylamphetamine and, less commonly, methcathinone; they are subject to certain trade controls under the provisions of UN1988 and subsequent domestic legislation (Appendix 5). Finally, ephedrine may be added to other powdered or tabletted drugs such as amphetamine or ketamine as an active diluent. Ephedrine and pseudoephedrine form part of a stereoisomeric quartet (Chapter 6).

In 1998, the WHO proposed that l-ephedrine and its racemate should be brought within the scope of UN1971. The separated d-isomer was not recommended for control as it is much less potent than the l-isomer.

---

[17] J. Hillebrand, D. Olszewski and R. Sedefov, *GHB and its precursor GBL: An emerging trend case study*, EMCDDA, 2008. http://www.emcdda.europa.eu/index.cfm?nNodeID = 7079

However, at a subsequent meeting of CND, the proposal was not accepted by the majority of UN signatories and was therefore not adopted.

More recently, with continued concern about the potential diversion of these substances towards methylamphetamine manufacture, the MHRA has produced a consultation document proposing that products containing more than 180 mg ephedrine or 720 mg pseudoephedrine should become Prescription Only Medicines (POM). The MHRA also proposed that both drugs should be controlled under the Misuse of Drugs Act. However, there is little evidence that these substances are intrinsically harmful; control would be seen as another way of restricting their use as precursors and would appear to conflict with the current listing of ephedrine and pseudoephedrine in the precursor chemical legislation (Appendix 5).

9.8 ADDITIONAL ANABOLIC STEROIDS

When anabolic steroids were first added to the Act in 1996 (followed by four androstenedione derivatives in 2003; Chapter 5), the candidate substances were largely those that were prohibited by the International Olympic Committee. The World Anti-Doping Agency (WADA) subsequently became the international body responsible for drug control in sport. By 2007, many more substances had been added to the WADA proscribed list. In 2008, ACMD recommended that 26 further substances should be controlled under the Misuse of Drugs Act as Class C drugs. None is covered by the generic definition (Chapter 6). They fall into four groups: 9 exogenous anabolic steroids; 1 endogenous anabolic steroid; 14 metabolites of anabolic steroids; and 2 nonsteroidal anabolic substances. The names shown in Table 9.7 are a BAN, an INN or an acceptable chemical name where no BAN or INN exist. Tetrahydrogestrinone (THG) and desoxymethyltestosterone ("Madol") are "designer steroids" that had been deliberately synthesised to circumvent WADA controls. The nonsteroidal substance zeranol has a structure based on a benztridecalactone system that can be configured loosely to resemble an oestrogenic steroid. Zilpaterol is an imidazobenzazepine with a β-hydroxy-phenethylamine chain nested within the ring system.

9.9 SUBSTANCES UNDER REVIEW BY WHO

A number of substances have been under discussion by the WHO as potential candidates for inclusion in UN1961 or UN1971. Apart from khat and ketamine, and a re-review of GHB (all of which are covered in

Table 9.7 Anabolic steroids and related substances recommended for control under the Misuse of Drugs Act in 2008.

Group 1: Exogenous steroids
1-Androstendiol
1-Androstendione
Boldione
Gestrinone
Danazol
Desoxymethyltestosterone
19-Norandrostenedione
Prostanozol
Tetrahydrogestrinone

Group 2: Endogenous steroids
Dihydrotestosterone

Group 3: Steroid metabolites
5α-Androstane-3α,17α-diol
5α-Androstane-3α,17β-diol
5α-Androstane-3β,17α-diol
5α-Androstane-3β,17β-diol
Androst-4-ene-3β,17β-diol
Androst-4-ene-3α,17α-diol
Androst-4-ene-3α,17β-diol
Androst-4-ene-3β,17α-diol
5-Androstenedione
Epidihydrotestosterone
3α-Hydroxy-5α-androstan-17-one
3β-Hydroxy-5α-androstan-17-one
19-Norandrosterone
19-Noretiocholanolone

Group 4: Nonsteroidal anabolic agents
Zeranol
Zilpaterol

other chapters), the list included zopiclone, tramadol, butorphanol and oripavine; these four are described below.

9.9.1 Zopiclone

Like zolpidem (Chapter 5), zopiclone ([8-(5-chloropyridin-2-yl)-7-oxo-2,5,8-triazabicyclo[4.3.0]nona-1,3,5-trien-9-yl] 4-methylpiperazine-1-carboxylate; Structure (9.17)) also acts on benzodiazepine receptors. Its pharmacological profile is similar to that of chlordiazepoxide. According to WHO, zopiclone leads to more reports of abuse than either nitrazepam or temazepam. In 2003, the ECDD recommended a critical review of zopiclone. But at the 34th meeting of ECDD in 2006, no recommendation for scheduling zopiclone was made.

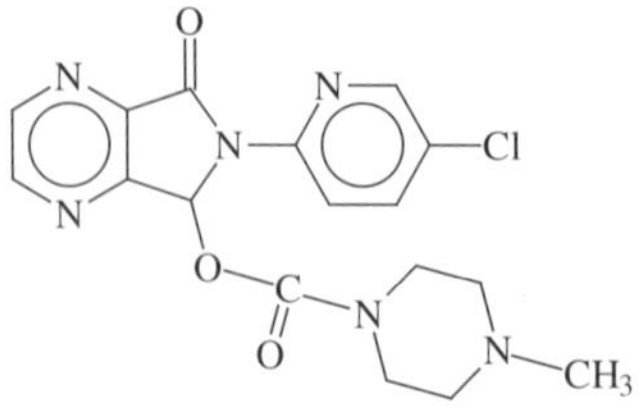

Structure (9.17) Zopiclone

9.9.2 Tramadol

Tramadol (1R, 2R-2-(dimethylaminomethyl)-1-(3-methoxyphenyl)-cyclohexanol; Structure (9.18)) is a synthetic substance with analgesic activity, and is used therapeutically in many countries. Some reports of abuse have likened tramadol to either codeine or dextropropoxyphene, but its abuse potential is considered to be less than that of buprenorphine or pentazocine. In 2003, the ECDD recommended that in view of the limited evidence of actual abuse, tramadol should be kept under surveillance. At the 34th meeting of ECDD in 2006, no recommendation for scheduling tramadol was made.

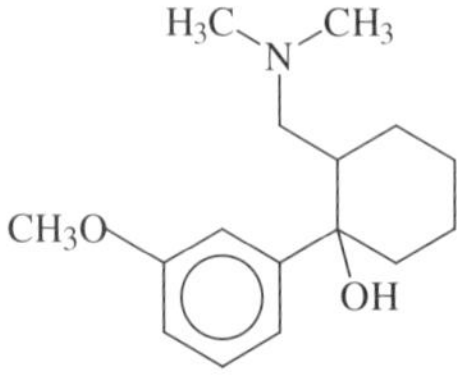

Structure (9.18) Tramadol

9.9.3 Butorphanol

Butorphanol (17-cyclobutylmethyl-morphinan-3,14-diol; Structure (9.19)) is a synthetic opioid with analgesic properties. It has a similar profile of activity to pentazocine – a Schedule III substance in UN1971. When administered parentally, 2–3 mg of butorphanol produce analgesia and respiratory depression equivalent to 10 mg morphine. There have been widespread reports of abuse of butorphanol; in 2003 the ECDD recommended that it should be critically reviewed. At the 34th meeting of ECDD in 2006, no recommendation for scheduling butorphanol was made.

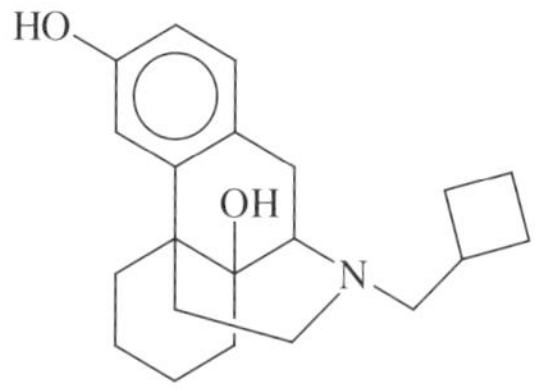

Structure (9.19) Butorphanol

9.9.4 Oripavine

Oripavine (6,7,8,14-tetrahydro-4,5-epoxy-6-methoxy-17-methyl morphinan-3-ol; Structure (9.20)) is a phenanthrene alkaloid otherwise known as O^3-desmethylthebaine. Concern over oripavine is based entirely on the fact that it is readily converted to thebaine by O-methylation. But although thebaine is listed in Schedule I of UN1961, abuse is almost unknown. In fact, thebaine was added to UN1961 because it is convertible into codeine and morphine. Thus, oripavine is a second-order precursor. It was considered by the WHO Expert Committee on Drug Dependence (ECDD) in 2003, but not recommended for critical review. However, at the 34th meeting of ECDD in 2006, WHO recommended that oripavine should be listed in Schedule 1 of UN1961. This was endorsed by a subsequent meeting of CND. It follows that oripavine will now need to be added to the Misuse of Drugs Act.

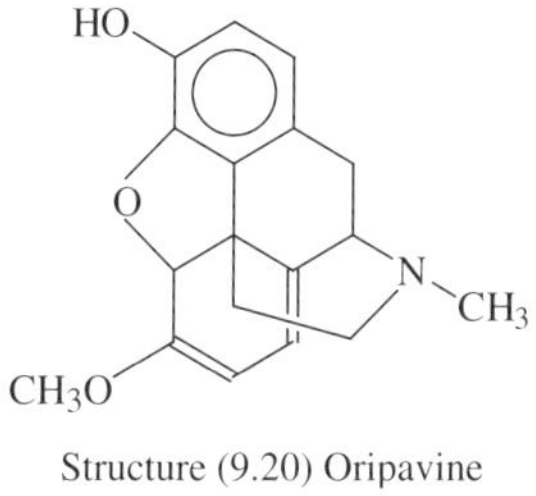

Structure (9.20) Oripavine

9.10 DIPHENYL-2-PYRROLIDINYLMETHANOL

Recently found in illicit tablets obtained from a UK website, (α,α-diphenyl-2-pyrrolidinylmethanol; D2PM, Structure (9.21)) is closely related to the Class C controlled drug pipradrol (α,α-diphenyl-2-piperidinemethanol; Structure (9.22)) and is thought to have similar CNS stimulant properties.

Structure (9.21) α, α-Diphenyl-2-pyrrolidinylmethanol

Structure (9.22) Pipradrol

9.11 METHYLHEXANEAMINE

Methylhexaneamine (2-methyl-4-aminohexane; Structure (9.23)) is a fairly obscure stimulant that was patented in 1944 and considered as an inhalant for nasal decongestion. In recent times, it has been marketed as a so-called dietary supplement for athletes under the unlicensed name "Geranamine".

Structure (9.23) Methylhexaneamine

9.12 MODAFINIL

Modafinil (2-(diphenylmethyl)sulfinylacetamide; Structure (9.24)) is one of the few CNS stimulants with therapeutic use that is not currently under international control. Claimed not to be a typical stimulant, but more a "wakefulness promoting agent", modafinil has been used in the treatment of narcolepsy, idiopathic hypersomnia and attention deficit hyperactivity disorder (ADHD). There have been few reports of its abuse. It has been suggested that it has cognitive enhancing and neuroprotective effects. It is a Schedule IV controlled substance in the US.

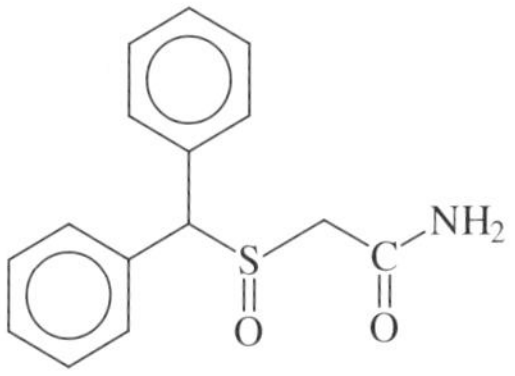

Structure (9.24) Modafinil

9.13 PHENAZEPAM

Phenazepam (10-bromo-2-(2-chlorophenyl)-3,6-diazabicyclo[5.4.0]undeca-2,8,10,12-tetraen-5-one; Structure (9.25)), also known as fenazepam, is produced in Russia, and appears to have a similar use and misuse profile to many other benzodiazepines. It has been reported on the illicit drugs market in Europe. Phenazepam is not listed in UN1971 and is not controlled by the Misuse of Drugs Act.

There are no current plans to create generic controls for the benzodiazepines. Although they have some structural similarity, even when brotizolam (a thienotriazolo-diazepine and not strictly a benzodiazepine) and clotiazepam (a thienodiazepine) are excluded, phenazepam and the remaining 32 benzodiazepines in the Act do not form a sufficiently homogeneous group amenable to a readily comprehensible group definition.

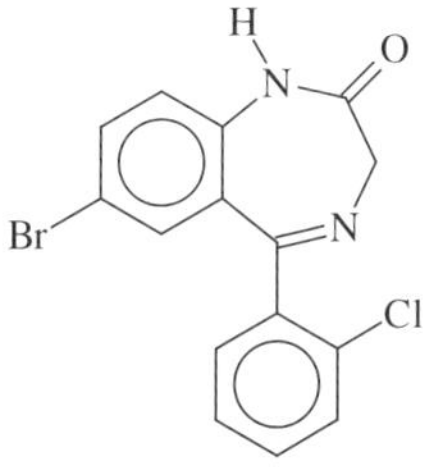

Structure (9.25) Phenazepam

9.14 MISCELLANEOUS OPIOIDS

A number of less-common opioids are occasionally misused. Nalbuphine, often in the form of the medicinal product Nubain® has been used by body-builders in the UK to overcome the pain of exercise.

As indicated in Chapter 6, dextromethorphan is currently excluded from control because of its value as a medicinal product (an anti-tussive) and low-misuse potential. This has not prevented the appearance of illicit tablets containing dextromethorphan.

More recently, glaucine (1,2,9,10-tetramethoxy-6-methyl-5,6,6a,7-tetrahydro-4H-dibenzo[de,g]quinoline; Structure (9.26)), a naturally occurring apomorphine derivative in the yellow horned poppy (*Glaucium flavum*) has been found in illicit tablets. Like dextromethorphan, glaucine is normally used as an anti-tussive in certain European countries (*e.g.* Glauvent® in Bulgaria). There have been reports of dissociative-type symptoms developing in patients using glaucine in illicit products[18]. Although studies suggest that glaucine has effects at central dopaminergic receptors, there is little other evidence to show that glaucine is psychoactive or likely to be widely abused.

Structure (9.26) Glaucine

Finally, lauroscholtzine, also known as californine, an opioid from the California poppy (*Eschscholtzia californica*) has been found in illicit tablets. Again, little is known of its pharmacology or abuse potential.

9.15 COGNITIVE ENHANCERS

Although hallucinogenic and psychedelic drugs are sometimes thought to "expand the mind", the search for substances that truly improve mental functions has proved more elusive. Sometimes known as nootropics[19] or simply "smart drugs", early candidates represented a diverse group of substances that included cholinergic drugs such as piracetam and its analogues, acetyl cholinesterase inhibitors, vitamins, amino acids, monoamine oxidase inhibitors, vasodilators and numerous herbal

---

[18] P.I. Dargan, J. Button, L. Hawkins, J.R.H. Archer, H. Ovaska, S. Lidder, J. Ramsey, D.W. Holt and D.M. Wood, *Detection of the pharmaceutical agent glaucine as a recreational drug*, Eur. J. Clin. Pharmacol., 2008, **64**, 553–554; http://www.springerlink.com/content/t01h728l60116104/
[19] http://en.wikipedia.org/wiki/Nootropic

products. Such drugs are claimed to change the availability of neurochemicals in the brain, to improve the oxygen supply to the brain, or to stimulate nerve growth. However, the efficacy of alleged nootropic substances in most cases has not been conclusively determined.

In the "Foresight Project on Brain Science, Addiction and Drugs: Drug Futures 2025", launched in 2005 by the former Department of Trade and Industry[20], one component was focused on cognition enhancers. It was believed that such substances would find use in the treatment of dementia, for those with specific cognitive impairment and those affected by the normal ageing process. A fourth group of users would be those who wished to use cognitive enhancers for non-therapeutic purposes. This in turn would raise social and ethical issues about whether and how such substances should be controlled by the criminal law. In a recent review[21] of brain science and addiction, it was noted that there were similarities in the future use of cognitive enhancers with the current use of performance enhancing drugs in sport.

9.16 MISCELLANEOUS NATURAL PRODUCTS CONTAINING PSYCHOACTIVE DRUGS

Table 7.1 lists the natural products named in the Misuse of Drugs Act 1971 or the Drugs Act 2005 (*i.e.* cannabis/cannabis resin, coca [leaves], opium, poppy-straw and certain fungi containing psilocin). A number of other plants contain controlled psychoactive substances, and these are discussed below. Table 9.8 shows examples of plants that contain non-controlled psychoactive substances. There is a general reluctance amongst drug legislatures to control more botanical entities. This is not a reflection of the fact that such substances are unusual or relatively harmless, but rather a recognition that such control would raise taxonomic difficulties. Insofar as it originates in an animal, the skin secretion of certain toads is an unusual natural product that contains a controlled drug, namely bufotenine. Whereas legal control of plants is one thing, control of an animal is quite another.

9.16.1 Peyote and Other Cacti

The peyote cactus (*Lophophora williamsii*) contains the hallucinogenic Class A controlled drug mescaline (3,4,5-trimethoxyphenethylamine),

---
[20] http://www.foresight.gov.uk/Previous_Projects/Brain_Science_Addiction_and_Drugs/index.html
[21] *Brain science, addiction and drugs*, An Academy of Medical Sciences working group report, 2008; http://www.acmedsci.ac.uk/p47prid47.html

Table 9.8 Botanical entities containing psychoactive substances not controlled in the UK or by the UN Conventions.

Botanical name	Common name	Principal component
Amanita sp.	Fly agaric	Muscimol[a]
Banisteriopsis caapi	Caapi	Harmine
Mitragyna speciosa	Kratom, Ketum, Biak	Mitragynine[b]
Myristica fragrans	Nutmeg	Myristicin
Piper methysticum	Kava Kava	Kavalactones[c]
Salvia divinorum	Mexican sage	Salvinorin A[d,e]
Tabernanthe iboga	Ibogaine	Ibogaine[f]

[a] http://en.wikipedia.org/wiki/Muscimol
[b] Mitragynine is controlled in Malaysia: K.B. Chan, C. Pakiam and R.A. Rahim, *Psychoactive plant abuse: the identification of mitragynine in ketum and in ketum preparations*, Bull. Narcotics, 2005, **57** (1–2), 249–256
[c] http://en.wikipedia.org/wiki/Kava
[d] C. Giroud, F. Felber, M. Augsburger, B. Horisberger, L. Rivier and P. Mangin, *Salvia divinorum: an hallucinogenic mint which might become a new recreational drug in Switzerland*, For. Sci. Int., 2000, **112**, 143–150
[e] *Salvia divinorum* is controlled in many European countries (Belgium, Denmark, Estonia, Finland, Germany, Italy, Norway, Spain and Sweden) and in some US States: http://druglaw.typepad.com/drug_law_blog/2008/03/salvia-four-sim.html
[f] Ibogaine is a Schedule I substance in the US Controlled Substances Act

but the intact plant is not controlled. Quite often, the separated outgrowths on the peyote cactus (mescal buttons) are seen. There have been several recent cases involving alleged production of mescaline from plant material. In the case of R-v-Sette, 2007, the defendant was charged with possessing 4.69 kg of a preparation containing mescaline with intent to supply. The preparation in question was described as "Peruvian Torch cactus" (believed to be *Trichocereus peruvianus*), which was in the form of dried vegetable matter. The prosecution accepted that cacti containing mescaline are not *per se* illegal, but argued that a *"preparation or other product"* of mescaline was illegal. The judge rejected the case on the grounds that the law was not sufficiently clear and that if it had been the intention to control certain cacti containing mescaline then this would have been included in the Drugs Act 2005 in the same way that fungi containing psilocin became controlled.

9.16.2 "Morning Glory" Seeds

Even though powdered seeds of "Morning Glory" (*Ipomoea* species) are part of the "Herbal High" repertoire, no prosecutions have been forthcoming for offences involving the preparation of a controlled drug, namely lysergamide. Hawaiian Baby Woodrose (*Argyreia nervosa*) also contains lysergamide.

9.16.3 Plants Containing Tryptamines

Many plants contain *N,N*-dimethyltryptamine (DMT) or related substances. Examples include *Diplopterys cabrerana*, *Psychotria viridis*, *Mimosa hostilis*[22] and species in the genera *Anadenanthera*. Again, there have been no prosecutions involving the extraction of the active substances.

9.17 CARISPRODOL

Carisoprodol (2-[[(Aminocarbonyl)oxy]methyl]-2-methylpentyl(1-methylethyl)carbamate; Structure (9.27)) is a centrally acting skeletal muscle relaxant. It was developed to create a drug with less abuse potential than meprobamate (Structure (9.28)), a metabolite of carisoprodol that is already a controlled substance (UN1971 Schedule IV, and Class C in the UK). There have been a number of case reports showing that carisoprodol also has abuse potential[23]. In late 2007, the European Medicines Agency (EMEA) recommended suspension of marketing authorisations for carisoprodol-containing medicinal products throughout the European Union. Carisoprodol is already a scheduled substance in some parts of the US.

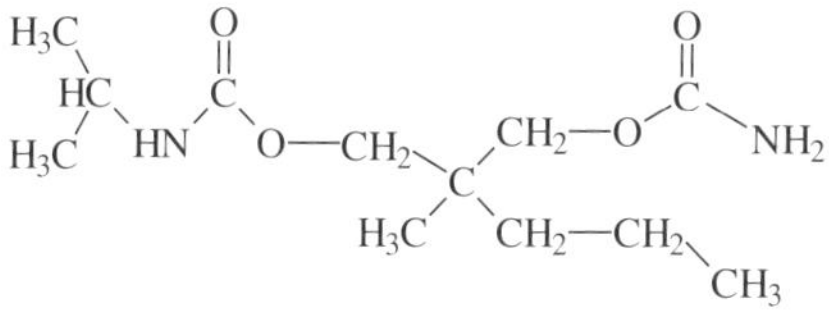

Structure (9.27) Carisoprodol

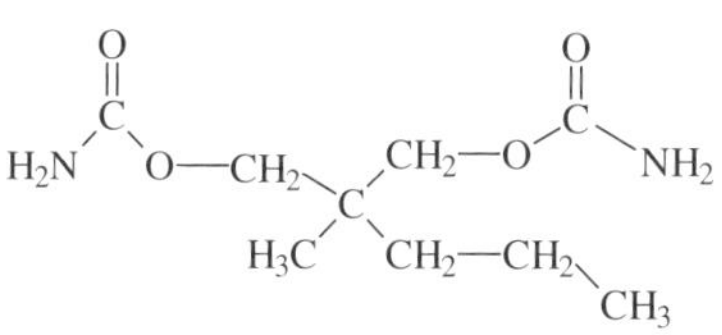

Structure (9.28) Meprobamate, a controlled metabolite of carisoprodol

---

[22] J.A. Fasanello and A.D. Placke, *The isolation, identification and quantitation of dimethyltryptamine (DMT) in Mimosa hostilis*, Microgram Journal, 2007, **5(1–4)**
[23] An early warning system in the Republic of Korea, based on the testing of urine and post-mortem specimens for noncontrolled substances, has provided evidence of the abuse of carisoprodal: H. Chung, *Role of drug testing as an early warning programme: the experience of the Republic of Korea*, Bull. Narcotics, 2005, **57(1–2)**, 231–248

9.18 MISCELLANEOUS HYPNOTICS AND OTHER SUBSTANCES

Three nonbenzodiazepine hypnotics: chloral hydrate, triclofos sodium and clomethiazole, are undoubtedly more toxic than the benzodiazepines. Clomethiazole in particular was responsible for many accidental and deliberate fatal poisonings in the 1980s. But, like the controlled hypnotic drug glutethimide, they are obsolescent, and their use continues to decline. Although chloral hydrate is controlled in the US, they are now unlikely candidates for inclusion in the Misuse of Drugs Act.

Misuse of benzydamine (3-(1-benzyl-1H-indazol-3-yloxy)-N,N-dimethylpropan-1-amine) has been reported in Poland[24]. This substance is a nonsteroidal anti-inflammatory drug with local anaesthetic properties. Normally intended for external use, it can produce hallucinations when large amounts are used orally.

9.19 CUTTING AGENTS AND ADULTERANTS

Many powdered drugs are diluted either before or after importation, and examples of diluents are shown in Appendix 12. It is often the case that large quantities may be seized in conjunction with controlled drugs at locations where adulteration, tabletting or encapsulation are taking place. Sometimes only cutting agents are recovered, and law-enforcement agencies may then use that as evidence of an association with criminal activities involving controlled drugs. The work of police and customs might be facilitated if a specific offence could be created of supplying such diluents. Although they are often widely available and have many other uses, there are parallels with drug precursors since, using the wording of the corresponding legislation, cutting agents could be described as "Substances useful for the manufacture of controlled drugs". However, unlike certain precursor chemicals, which may have little use other than to manufacture a specific drug, cutting agents are often ubiquitous materials. If person A supplies B with, for example, glucose then it might be difficult to prove that A knew or suspected what B would do with it. The law relating to complicity in crime is complex, and depends on whether the anticipated crime is committed or not. An answer might be found in the Serious Crime Act 2007. Although not yet in force, this Act creates offences of encouraging or assisting in the commission of a crime such that a supplier of a cutting agent might be held liable even if the anticipated offence was not committed.

---

[24] J.S. Anand, M. Lukasik-Glebocka and R.P. Korolkiewicz, *Recreational abuse with benzydamine hydrochloride (tantum rosa)*, Clin. Toxicol., 2007, **45**, 103–104

CHAPTER 10
Generic and Analogue Control – International Comparisons

10.1 GENERIC DEFINITIONS IN NEW ZEALAND

While most countries have chosen to implement only the essential elements required in international law by the 1961 and 1971 United Nations Convention, a few have extended the scope of their legislation to a wider range of substances or have introduced generic or analogue control. One example is the Irish Republic, where the Misuse of Drugs Act 1975 closely follows the UK approach. The Misuse of Drugs Act 1977 of New Zealand also has a system of generic definitions, but they are less closely related to the British model. The definitions cover analogues of amphetamine, pethidine, phencyclidine, fentanyl, methaqualone and dimethyltryptamine[1]. These six families were considered to be the primary focus of designer drugs (Chapter 6) in the 1980s. Three of them: amphetamine; methaqualone; and the tryptamines are described below. Although the New Zealand legislation is similar to the UK model in dividing scheduled drugs between three classes: A, B and C, there are some major differences. Apart from the scope and detail of the generic definitions, they are set out in Part VII of the Third Schedule of that Act alongside Class C drugs. In other words, the structural variants are automatically assigned to Class C even if the parent (*e.g.* pethidine, fentanyl) is in Class A or Class B. A second major difference is that a

---

[1] G.J. Sutherland, "*Designer" drugs legislation in New Zealand and elsewhere*, Analog: Australian Forensic Drug Analysis Bulletin, 1988, **10(3)**

less-specific analogue control sits alongside the generic control. This, and the corresponding US analogue control, is described below. The New Zealand generic controls have been adopted into legislation in Australia.

10.1.1 Amphetamine Derivatives

The full definition reads as follows:

"Amphetamine analogues, in which the 1-amino-2-phenylethane nucleus carries any of the following radicals, either alone or in combination:

(a) *1 or 2 alkyl radicals, each with up to 6 carbon atoms, attached to the nitrogen atom;*
(b) *1 or 2 methyl radicals, or an ethyl radical, attached to the carbon atom adjacent to the nitrogen atom;*
(c) *a hydroxy radical, attached to the carbon atom adjacent to the benzene ring;*
(d) *any combination of up to 5 alkyl radicals and/or alkoxy radicals and/or alkylamino radicals (each with up to 6 carbon atoms, including cyclic radicals) and/or halogen radicals and/or nitro radicals and/or amino radicals, attached to the benzene ring."*

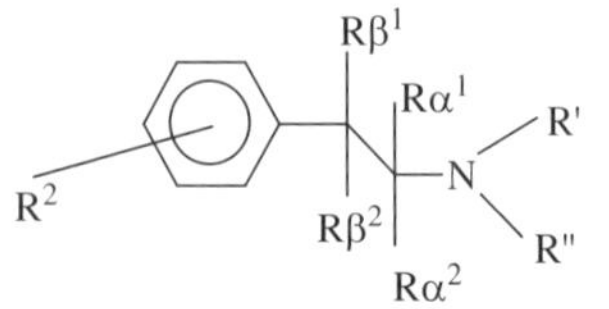

Structure (10.1) Phenethylamine showing substitution patterns

To qualify as a Class C controlled drug in New Zealand, the following criteria in Structure (10.1) must be satisfied:

$R' = H$ or alkyl (not more than six carbon atoms)

$R'' = H$ or alkyl (not more than six carbon atoms)

$R^{\alpha 1} = H$ or methyl

$R^{\alpha 2} = H$, methyl or ethyl

$R^{\beta 1} = H$ or OH

$R^{\beta 2} = H$

R^2 = alkyl, alkoxy, alkylamino, halogen, nitro or amino (either singly or in any combination) or a cyclic group, where no substituent has more than six carbon atoms.

Although these rules are broader in scope than those in the UK legislation (Chapter 6), in practical terms they do not capture many more of those compounds that have been found in illicit preparations.

10.1.2 Methaqualone Derivatives

The full definition reads as follows:

> "Methaqualone analogues, in which the 3-arylquinazolin-4-one nucleus has additional radicals, either alone or in combination, attached as follows:
>
> (a) an alkyl radical, with up to 6 carbon atoms, attached at the 2 position;
> (b) any combination of up to 5 alkyl radicals and/or alkoxy radicals (each with up to 6 carbon atoms, including cyclic radicals) and/or halogen radicals, attached to each of the aryl rings."

Structure (10.2) The methaqualone nucleus showing substitution patterns

To qualify as a Class C controlled drug in New Zealand, the following criteria in Structure (10.2) must be satisfied:

R^1 = alkyl (not more than six carbon atoms)

R^2 = any combination of up to five substituents that include alkyl, alkoxy, halogen or a cyclic group, where no substituent has more than six carbon atoms.

There is no corresponding generic control in the UK legislation; methaqualone and its derivatives were never common, and today are almost unknown. But three analogues of methaqualone had been seen in Europe. Mecloqualone, which is captured by the above rules, is a Schedule II substance under UN1971, and was added to the UK Misuse of Drugs Act in 1984. Methylmethaqualone and the brominated analogue, mebroqualone, both of which are covered by the above rules, were found in Germany in 1997.

10.1.3 Tryptamine Derivatives

The full definition reads as follows:

> "*DMT (dimethyltryptamine) analogues, in which the 3-(2-aminoethyl)indole nucleus has additional radicals, either alone or in combination, attached as follows:*
>
> (a) *1 or 2 alkyl radicals, each with up to 6 carbon atoms, including cyclic radicals, attached to the amino nitrogen atom;*
> (b) *1 or 2 methyl groups, or an ethyl group, attached to the carbon atom adjacent to the amino nitrogen atom;*
> (c) *any combination of up to 5 alkyl radicals and/or alkoxy radicals (each with up to 6 carbon atoms, including cyclic radicals) and/or halogen radicals, attached to the benzene ring.*"

Structure (10.3) Tryptamine showing substitution patterns

To qualify as a Class C controlled drug in New Zealand, the following criteria in Structure (10.3) must be satisfied:

R' = methyl or other alkyl including [R' to R''] cyclic groups

R'' = methyl or other alkyl. For R' and R'', not more than six carbon atoms

$R^{\alpha 1}$ = H or methyl

$R^{\alpha 2}$ = H, methyl or ethyl

$R^1 = R^2 = R^{\beta 1} = R^{\beta 2}$ = H

R^4, R^5, R^6 and R^7 = any combination of up to five substituents that include alkyl, alkoxy, halogen or a cyclic group, where no substituent has more than six carbon atoms.

In contrast to the UK definition of a substituted tryptamine, the above rules exclude some of the important ring-hydroxy tryptamines (*e.g.* psilocin), but on the other hand do cover substances such as α-ethyltryptamine.

10.2 DRUG "ANALOGUES"

The general principles of generic, *i.e.* structure-specific, controls have been discussed both for the UK (Chapter 6) and for New Zealand (see above). A much broader type of analogue control can be found in a few countries. Examples from the US and New Zealand are described below.

10.2.1 US Analogue Control

The Controlled Substances Analogue Enforcement Act of 1986 defines analogues in the following way:

"Controlled substance analogue means a substance –

(i) *the chemical structure of which is substantially similar to the chemical structure of a controlled substance in schedule I or II.*
(ii) *which has a stimulant, depressant, or hallucinogenic effect on the central nervous system that is substantially similar to or greater than the stimulant, depressant, or hallucinogenic effect on the central nervous system of a controlled substance in schedule I or II; or*
(iii) *with respect to a particular person, a substance which such person represents or intends to have a stimulant, depressant, or hallucinogenic effect on the central nervous system substantially similar to or greater than the stimulant, depressant, or hallucinogenic effect of a controlled substance in schedule I or II."*

In an appeal heard in 1996 (United States v. Allen McKinney), the "Analogue Act" was deemed not to be constitutionally vague[2]. The case concerned sale of aminorex before it became explicitly controlled, and the sale of phenethylamine as a substitute for meth(yl)amphetamine. Despite this ruling, there is a widespread view in Europe that analogue controls are less satisfactory from a legal viewpoint. Whereas with explicit listing of substances in a schedule or even a generic definition, the status of a substance is clear from the outset, the use of analogue legislation requires that a court process should determine whether the substance is or is not controlled and hence whether any offence has been committed. This might be seen as a cumbersome method, requiring as it does expert chemical and pharmacological testimony in every case. However, it cannot be denied that from a US perspective The Controlled Substances Analogue Enforcement Act of 1986 has been highly successful in curtailing the proliferation of designer drugs. The US government has prosecuted a substantial number of individuals for the manufacture and distribution of analogues of MDA, amphetamine, pethidine (meperidine), fentanyl and others. The US view is that most of the substances in PIHKAL could meet the definition of a controlled analogue.

10.2.2 New Zealand Analogue Control

In New Zealand, the Misuse of Drugs Amendment Act (No. 2) 1987 introduced the definition of a "Controlled Drug Analogue" as "*any substance, such as the substances specified or described in Part VII of the Third Schedule to this Act, that has a structure substantially similar to that of any controlled drug; . . .* ". The Amendment goes on to exclude any substance listed elsewhere in the Misuse of Drugs Act or as a pharmacy-only medicine, restricted medicine or prescription medicine under the Medicines Act and Regulations. To a certain extent, this Amendment was inspired by, and modelled on, the US analogue controls.

The application of the analogue provisions of the Amendment Act is not limited to the families of substances listed in Part VII of the Third Schedule (*i.e.* amphetamine, pethidine, phencyclidine, fentanyl, methaqualone and dimethyltryptamine). But, the definition of what constitutes "substantially similar" is a potentially arguable issue for

---

[2] Anon. *U.S. Analogue statute ruled not constitutionally vague*, Clandestine Laboratory Investigating Chemists Association, 1996, **6(4)**, 5–6

substances other than those six categories, and thus far there is minimal case law to clarify this.

10.3 EMERGENCY SCHEDULING PROVISIONS

In Chapter 6, it was noted how, in the UK, the appearance of "designer drugs" in the late 1970s and early 1980s led to the introduction of generic controls. As described above, these were later introduced by some other countries, while others adopted the analogue approach. However, in the US, the first reaction was to introduce a scheme in 1984 whereby a substance could be temporarily added to Schedule 1 of the Controlled Substances Act for a period of one year. The conditions that had to be satisfied were that the substance presented an imminent hazard to the public safety, and that it wasn't already listed in another Schedule of the Act[3]. This temporary measure could be extended by six months provided, by then, procedures had been initiated to control the substance permanently. There was still a requirement on the authorities to provide some evaluation of the abuse potential of the substance, even if these had to be inferred from structure–activity relationships and comparison with similar compounds. Within a few years, it was recognised that, while a valuable tool, emergency scheduling was not in itself enough to limit the illicit manufacture of designer drugs. This need for a more proactive stance gave rise to the Controlled Substances Analogue Enforcement Act of 1986 (see above). To a certain extent, analogue control reduced the need for emergency scheduling, but it is still used in the US; recent examples included the temporary listing of 1-benzylpiperazine (BZP) and 1-(3-trifluoromethyl-phenyl)piperazine (TFMPP) in 2002. Subsequently, BZP was made subject to permanent control (Schedule I) while TFMPP was removed from control because of a lack of evidence of harmful properties.

In New Zealand a somewhat different approach to emergency legislation has been used. Although the classification system has three nominal classes, *i.e.* A, B and C, a fourth category, Class D, was set up to deal with BZP. Only limited controls were placed on BZP, such as a prohibition of sale to minors. However, following a further review, BZP together with *m*CPP, TFMPP, *p*FPP, MBZP and MeOPP became Class C controlled drugs[4] on 1st April 2008.

---

[3] http://www.deadiversion.usdoj.gov/fed_regs/sched_actions/2002/fr07182.htm
[4] Misuse of Drugs (Classification of BZP) Amendment Bill; http://www.parliament.nz/en-NZ/PB/Legislation/Bills/d/3/d/00DBHOH_BILL8220_1-Misuse-of-Drugs-Classification-of-BZP-Amendment.htm

CHAPTER 11
The Drug Classification Debate

11.1 INTRODUCTION

The important features of the Misuse of Drugs Act 1971 were described in Chapter 5. There appear to be few documentary records of how the UK came to adopt the two-dimensional approach to drug control, *i.e.* the concept that the legal status of every controlled substance was defined by both a Schedule in the Misuse of Drugs Regulations and a Class in the Act. In a statement by the Home Secretary (James Callaghan) in 1970, the purpose of introducing drug classes was: *" . . . to make, so far as is possible, a more sensible differentiation between drugs. It will divide them according to their dangers and harmfulness in the light of current knowledge and it will provide for changes to be made in the classification in the light of new scientific knowledge"*. There is no clear reason why a three-Class system was adopted. Anecdotal accounts from the late 1960s and early 1970s, when the Misuse of Drugs Bill was being debated, suggest that the Government's plan was to have a two-Class approach. Although the terms are now largely obsolete, the idea may have arisen because, at the time, there was a commonly held view that drugs of abuse could be divided into "hard drugs" and "soft drugs". However, the question of where to place cannabis in this structure is said to have caused so much debate that a compromise was reached whereby cannabis became Class B and the "soft" substances were distributed between Class B and a new Class C. The problem of cannabis can be associated with the aphorism that *"Everything starts and finishes with cannabis"*. Whatever the truth, when the Misuse of Drugs Act came in

---

Forensic Chemistry of Substance Misuse: A Guide to Drug Control
By L.A. King
© L.A. King 2009
Published by the Royal Society of Chemistry, www.rsc.org

force, Class B contained only 13 entries and Class C contained 10. By contrast, there were over 90 substances in Class A. It should be noted that the classification system is entirely independent of scheduling under the 1961 and 1971 United Nations Conventions; movements between classes is a purely domestic issue.

11.2 THE 1979 REVIEW BY ACMD

The first review of drug classification in the Misuse of Drugs Act was published in 1979[1]. It seems that the Advisory Council on the Misuse of Drugs (ACMD) was broadly satisfied with the overall classification system since few recommendations for change were made. Reclassification of methaqualone from Class C to Class B was accepted by the Government, but moving cannabis and cannabis resin from Class B to Class C was not.

11.3 THE INDEPENDENT ENQUIRY INTO THE MISUSE OF DRUGS ACT (2000)

In the twenty years that followed the 1979 review, many more substances were added to the Act and the generic controls were extended. But few questions were raised about classification. In the absence of any further scrutiny by ACMD, the Police Foundation, a body independent of Government, decided to commission its own review of the Act in the late 1990s. This became the Independent Enquiry into the Misuse of Drugs Act with a remit to examine the changes that had taken place in society in the 30 years since the Act appeared and to ask whether the legislation needed to be revised to make it both more effective and more responsive to those changes. Their report[2] was published in 2000; it made many recommendations covering enforcement, offences, treatment and research. For the classification system, it recommended that the three-Class approach should be retained and that there should be clear criteria for additions to and transfers between the classes. Apart from cannabis and cannabis resin, which should be transferred from Class B to Class C, the Independent Enquiry also recommended that a number of other controlled drugs should be reclassified. It proposed that

---
[1] Advisory Council on the Misuse of Drugs, *"Report on a Review of the Classification of Controlled Drugs and of Penalties under Schedule 2 and 4 of the Misuse of Drugs Act 1971"*, Home Office, 1979
[2] *Drugs and the Law: Report of the Independent Inquiry into the Misuse of Drugs Act 197* – see Bibliography

ecstasy and related compounds should be moved from Class A to B, LSD from Class A to B, and buprenorphine from Class C to B.

However, the Government of the day did not accept any of the recommendations concerning drug reclassification. Although it would be included in risk-assessment exercises carried out by ACMD after 2000, there has been little subsequent debate about the status of LSD. This may be partly a reflection that its use has become uncommon. Buprenorphine is an opioid analgesic used clinically as a pre-medication and an adjunct to anaesthesia as well as in the treatment of drug dependence. It was considered by ACMD in 2002 and again in 2005, and reclassification (to Class B) was supported. In the light of a review by ACMD in 2006, information emerged that buprenorphine was increasingly used as an alternative to methadone in treating drug dependence, as well as an analgesic in veterinary medicine. In the meantime, buprenorphine had also been considered by WHO/ECDD, whereby no change had been recommended in its status under UN1971. There was little evidence that it was being more widely abused in the UK, and therefore the original proposal for reclassification was abandoned. Thus, buprenorphine remains in Class C.

11.4 RECLASSIFICATION OF CANNABIS (2001–4)

In late 2001, the Home Secretary (David Blunkett), giving evidence to the Parliamentary Home Affairs Committee, announced that the Government was proposing that cannabis and cannabis resin should be reclassified as Class C drugs. In large measure, this proposal was prompted by the amount of police time spent processing relatively minor cannabis offences, when they were expected to focus enforcement efforts on Class A drugs. It was also recognised that police activity against cannabis users was antagonising many young people. As would happen again in later years, the political proposal often appeared to pre-empt a recommendation by ACMD. But reclassification of cannabis, cannabis resin and the "cannabinols" was supported by ACMD; their report[3] was published in March 2002. There was no intention that cannabis, cannabis resin or the "cannabinols" should be rescheduled with respect to the Regulations. In his speech in 2001, David Blunkett had stressed that reclassification would not amount to legalisation or decriminalisation,

---

[3] Advisory Council on the Misuse of Drugs, *The Classification of Cannabis under the Misuse of Drugs Act 1971* – see Bibliography

but later events showed that large sections of the public or the media had difficulty in distinguishing these concepts from reclassification.

It is apparent that the initial enthusiasm for reclassification became tempered in 2003. Firstly, following the ACMD report of March 2002, a Modification Order was not published until 2003 (S.I.3201), and this did not come into force until 29 January 2004. Secondly, and much more significantly, the penalties for certain offences involving Class C drugs were changed by the Criminal Justice Act 2003. This allowed the power of arrest to be used for the possession of cannabis, whereas before re-classification the possession of a Class C drug was not an arrestable offence. The maximum prison sentence for supplying any Class C drug was increased from 5 years to 14 years, *i.e.* similar to the penalty associated with Class B drugs. These measures were seen as immediately negating part of the impact of reclassification. Although the maximum prison sentence for possessing any Class C drug was simultaneously reduced from 5 years to 2 years, there had never been many instances of imprisonment for simple possession of any Class C drug.

11.5 HOME AFFAIRS SELECT COMMITTEE (2001–2)

The Home Affairs Select Committee is appointed by the House of Commons to examine the expenditure, administration and policy of the Home Office and the Lord Chancellor's Department (now part of the Ministry of Justice) and certain other public bodies. The Committee chooses its own areas of investigation. For its third report of the session 2001–2, it investigated drugs policy. The Committee's report[4] was published in May 2002. This wide-ranging inquiry made two recommendations about drugs classification. Firstly, it supported the Home Secretary's proposal of 2001 to reclassify cannabis from Class B to Class C. Secondly, and echoing the findings of the Independent Enquiry into the Misuse of Drugs Act, it recommended that ecstasy (MDMA) should move from Class A to Class B. The latter was again rejected by the Government largely on the alleged fatal toxicity of MDMA. As would be repeated on later occasions, it was stated that: *"The Government has no intention of reclassifying ecstasy. Ecstasy can and does kill unpredictably; there is no such thing as a "safe dose". The Government firmly believes that ecstasy should remain a Class A drug."* This focus on the fatal toxicity of ecstasy reflected public concern following a number

---

[4] *The Government's Drugs Policy: Is it Working?* – see Bibliography

of high profile deaths, most notably that of a young woman named Leah Betts.

11.6 REVIEW OF CANNABIS (2005–6)

In early 2005, the Home Secretary (Charles Clarke) asked the ACMD to consider whether it had changed its position from that set out in its March 2002 report in the light of new evidence that associated cannabis with mental health problems and the prevalence of cannabis with high levels of THC. Following a review by ACMD, its report[5] was published in January 2006. The report supported the retention of cannabis and "cannabinols" in Class C.

11.7 HOME OFFICE PROPOSALS FOR REVIEWING THE CLASSIFICATION SYSTEM (2006)

The Government's somewhat reluctant decision to accept the recommendations of the 2005 report on cannabis from ACMD gave rise in January 2006 to a proposal by the Home Secretary (Charles Clarke) that there should be a review of the entire drug classification system. It was felt that the system had been in operation for 35 years, had never been fully reviewed and was not completely understood by the public. It was believed, at the time, that the Government favoured a two-Class system, an idea that had been originally advocated in the late 1960s. However, later that same year, Charles Clarke left his post as Home Secretary. His replacement, John Reid, quickly abandoned the proposed review of drug classifications citing lack of evidence amongst the major stakeholders that this was a high priority issue.

11.8 METHYLAMPHETAMINE (2006)

The pharmacological properties of methylamphetamine are similar to those of amphetamine, and it is likely that a drug user would be unable to distinguish them when administered in the same way. Despite being prevalent in the US and certain countries of the Far East, methylamphetamine is uncommon in the UK and most of Western Europe. The

---

[5] Advisory Council on the Misuse of Drugs, *Further consideration of the classification of cannabis under the Misuse of Drugs Act 1971* – see Bibliography

much higher prevalence of the drug in the US has encouraged numerous commentators over many years to suggest that Europe is overdue for a rapid rise in consumption, rather as occurred with crack cocaine nearly twenty years ago. There are good reasons for thinking that methylamphetamine will not become more widespread as long as amphetamine is readily and cheaply available: a situation that is not true in the US.

What little drug is seized by law-enforcement agencies nearly always turns out to be powders or tablets, both of which would normally be ingested. It is often the case that material claimed to be "crystal meth" is found, on analysis, to be crystalline MDMA. Methylamphetamine hydrochloride is sufficiently volatile that it can be smoked, although this practice would normally only occur with the pure material (sometimes known as "Ice") to avoid the undesirable combustion of cutting agents. Since "Ice" is almost never seen in Europe, it can be reasonably assumed that smoking methylamphetamine is an unusual activity. But there is little doubt that smoking this drug is a much more harmful activity than ingestion. Drugs that are smoked (*e.g.* tobacco/nicotine, heroin, crack cocaine, cannabis/THC) reach the brain far more quickly than when ingested. As a consequence their addictive potential is higher.

Methylamphetamine had not been included in the original ACMD risk-assessment survey (see below). However, following a detailed review carried out by ACMD[6], and using the same parameters of harm in a risk-assessment exercise, methylamphetamine scored 2.12 out of a possible 3. This was higher than had been given to amphetamine (score = 1.66), and the third highest score of twenty-two substances. The initial recommendation was that no change in the law was needed. However, the situation was kept under review and, following pressure from law-enforcement agencies, a recommendation for reclassification was made. Methylamphetamine became a Class A drug in 2006. The new evidence presented largely revolved around an apparent increase in attempts to manufacture the drug. In reality, little has changed and there have been no significant seizures of clandestine laboratories. Recent surveys show that not only does it account for less than 1% of amphetamine seizures, but that it is extremely rare in urine samples tested under employee drug testing regimes and rarely features in calls for advice to poison control centres[7].

---

[6] Advisory Council on the Misuse of Drugs, *Methylamphetamine Review*, 2005; http://www.drugs.gov.uk/publication-search/acmd/ACMD-Meth-Report-November-2005?view = Binary

[7] D.M. Wood, J. Button, T. Ashraf, S. Walker, S.L. Green, N. Drake, J. Ramsey, D.W. Holt and P.I. Dargan, *What evidence is there that the UK should tackle the potential emerging threat of methamphetamine toxicity rather than established recreational drugs such as MDMA ("ecstasy")?*, Quart. J. Med. Advance Access published January 25, 2008

11.9 SELECT COMMITTEE ON SCIENCE AND TECHNOLOGY (2006)

As part of a wider investigation into the provision of expert advice to Government, the Parliamentary Select Committee on Science and Technology examined the drug classification system and the role played by ACMD. In the Committee's report[8], a number of deficiencies were identified. There was specific criticism of the activities of the ACMD including a lack of transparency and for apparently muddled thinking in its decision to reclassify methylamphetamine from Class B to Class A (see above) so soon after its own report recommended that no change of status was required. The classification of drugs was considered to be generally inconsistent, and the Government was criticised for using the ABC system to send a signal to users and society at large, that was at odds with the stated objective of classifying drugs on the basis of harm. The Government was criticised for using the Drugs Act 2005 to make fresh magic mushrooms a Class A drug: a mechanism that contravened the spirit of the Misuse of Drugs Act by giving ACMD no formal opportunity to consider the evidence. There was little evidence that classification had a deterrent effect, and the system was described as "*not fit for purpose*". In the view of the Committee, it should be replaced with a scientifically based scale of harm decoupled from penalties. The report concluded that there should be a thorough review of the current system as proposed by the former Home Secretary in early 2006. As an annex to their report, the Committee included an early version of a publication on a scale of drug harm, which is described in more detail below. Among specific points, the Select Committee recommended that there should be an urgent review of the legal status of ecstasy.

11.10 ROYAL SOCIETY OF ARTS REPORT (2007)

Although making no specific suggestions regarding reclassification, the report[9] of the Royal Society of Arts Commission on Illegal Drugs, Communities and Public Policy provided a critique of the Misuse of Drugs Act, arguments for and against legalisation and options for change. However, the latter were largely influenced by the recently completed report from the Parliamentary Select Committee on Science and Technology and the subsequently published Scale of Drug Harm (see below). But even without anything particularly novel arising from

---

[8] *Drug classification: making a hash of it?* – see Bibliography
[9] *Drugs: Facing Facts* – see Bibliography

this review, it says much about the state of drug policy in the UK in the early years of the 21st century that the Royal Society of Arts felt there was a need to add yet a further voice to a debate already crowded with reviews.

11.11 SCALE OF DRUG HARM (2007)

In 2000, ACMD started work on risk assessments of what would eventually amount to twenty substances, spanning all three Classes (A, B and C) and some noncontrolled substances. This large project partly had its origins in the risk assessments that were carried out for the Independent Enquiry into the Misuse of Drugs Act (see earlier), but was also prompted by progress being made at European level by the EMCDDA in defining a systematic approach to the evaluation of the harms of "new synthetic drugs" (Chapter 4). Twenty substances were ranked on a scale of harm, based on a nine-parameter matrix that included both personal and social harms[10]. Figure 11.1 shows the overall scores for the twenty substances together with methylamphetamine (score = 2.12) and 1-benzylpiperazine (score = 0.86). The current status of the substances under the Misuse of Drugs Act is also shown in Figure 11.1. Of the twenty-two substances, fourteen had been previously scored by a panel of psychiatrists on a similar matrix during the Independent Inquiry into the Misuse of Drugs Act (see Bibliography), although those scores had not been published at the time. Good agreement (Pearson's correlation coefficient: $r = 0.892$; $n = 14$; $P < 0.001$) was reached between the ranking obtained by ACMD members and the psychiatrists (Figure 11.2).

Alcohol (score = 1.85 out of a possible 3) was ranked the highest of noncontrolled drugs. Although there is no suggestion that it could ever be controlled by the Misuse of Drugs Act, alcohol was deemed to be as harmful as many substances in Class A.

The next highest noncontrolled drug was ketamine (score = 1.75 out of a possible 3). This led to a review of ketamine by ACMD[11], following which a recommendation was made that it should be controlled; in 2005 it was listed in Class C. It could be argued from its score that ketamine should have been placed in Class B, but to a large extent the decision was precipitated by a need to control importation of nonmedicinal products

---

[10] D. Nutt, L.A. King, W. Saulsbury and C. Blakemore, *Developing a rational scale for assessing the risks of drugs of potential misuse*, Lancet, 2007, **369**, 1047–1053
[11] http://drugs.homeoffice.gov.uk/publication-search/acmd/ketamine-report.pdf

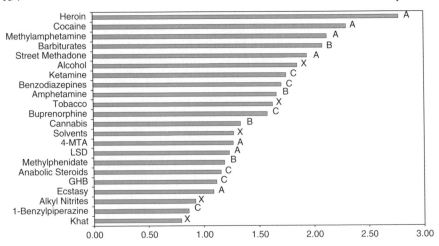

Figure 11.1 Overall harm scores of twenty-two substances examined by a group of independent experts (ACMD). The respective classification (A, B or C) under the Misuse of Drugs Act (X = unclassified) is shown against each bar.

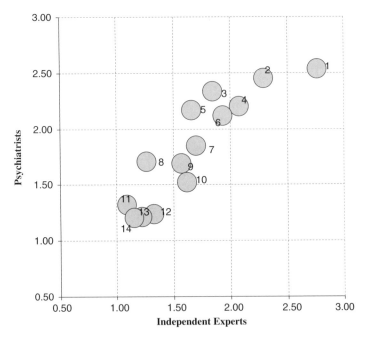

Figure 11.2 Overall harm scores of fourteen substances by a panel of independent experts (ACMD) and a group of psychiatrists. Key: 1 = heroin; 2 = cocaine; 3 = alcohol; 4 = barbiturates; 5 = amphetamine; 6 = methadone; 7 = benzodiazepines; 8 = solvents; 9 = buprenorphine; 10 = tobacco; 11 = ecstasy; 12 = cannabis; 13 = LSD; 14 = steroids.

containing ketamine; there was little enthusiasm for wishing to criminalise users. Although it had first come to notice as a problem in the UK as early as 1992, no convincing evidence emerged in the intervening years that it should have been controlled sooner. Thus, despite the high score achieved by ketamine, it might have seemed illogical for it to be immediately classified higher than Class C.

Khat achieved the lowest score (0.80 out of a possible 3). In line with this finding, a subsequent review by ACMD[12] recommended that khat should remain outside the Act. The alkyl nitrites (score = 0.92) are also unlikely substances for control under the Act. Amongst the controlled drugs in Figures 11.1 and 11.2, the placement of heroin and cocaine in the first and second places is of little surprise. However, the ranking of some seems misplaced, and the following conclusions might be drawn:

- Ecstasy and LSD should be reclassified from Class A to Class B
- Barbiturates should be reclassified from Class B to Class A
- Benzodiazepines should be reclassified from Class C to Class B

11.12 REVIEW OF CANNABIS (2007–8)

The year 2007 brought a new Home Secretary (Jacqui Smith) and yet another call for ACMD to reinvestigate the status of cannabis. As has happened before with initiatives in the drug-control field, this latest move was not entirely unconnected with wider political imperatives. On this occasion, it was the Prime Minister (Gordon Brown) who chose to announce the need for the review, and again pre-empted the outcome of the review by claiming that cannabis should be moved to Class B. The review took place in early 2008. Although there had been a continued decline in the population prevalence of cannabis use, there was a growing concern about its harmfulness. This centred around the apparently strengthening evidence that cannabis was a causative agent for chronic psychosis, particularly schizophrenia, the fact that high-potency forms (sinsemilla/skunk) were now the dominant market product, increasing evidence of dependence and the social harms caused by the recent and rapid proliferation of large indoor cannabis farms and their association with organised crime. The recommendation of ACMD[13] was that cannabis should stay in Class C, but in May 2008 the Government

---

[12] http://drugs.homeoffice.gov.uk/publication-search/acmd/khat-report-2005/
[13] Advisory Council on the Misuse of Drugs, *Classification of cannabis and public health* – see Bibliography

decided that "cannabis products" should be reclassified to Class B. It is planned[14] that this will take effect in 2009. In ignoring the advice of the ACMD, the Home Secretary used the precautionary principle and admitted *"erring on the side of caution"* in relation to the possible harmful effects of cannabis on the mental health of future generations. Together with reclassification, the Home Secretary also announced other measures to tackle cannabis misuse which included: more robust enforcement against cannabis supply and possession – those repeatedly caught with the drug will not just receive warnings; a new strategic and targeted approach to tackling cannabis farms and the organised criminals behind them; introducing additional aggravating sentencing factors for those caught supplying cannabis and other illegal substances near further and higher educational establishments, mental health institutions and prisons; and working with the Association of Chief Police Officers to look at how existing legislation and powers can be used to curtail the sale and promotion of cannabis paraphernalia.

This latest re-examination of cannabis is likely to be the last for some time. Although one of the recommendations of the 2008 ACMD report was that there should be a further review in 2010, it would seem doubtful that there could be any political will for this to happen. Furthermore, it seems that high-potency cannabis will remain the dominant product for some time. It is also improbable that our understanding of the harms caused by cannabis, particularly harms to mental health, will be further increased by yet more epidemiological studies. Since 1979, when ACMD first recommended that cannabis should be moved to Class C – a period when our knowledge about the effects of and use of cannabis have increased dramatically – the ACMD has been remarkably consistent in its view on classification. Table 11.1 summarises the various recommendations for the classification of cannabis that have been made over nearly 40 years.

11.13 REVIEW OF ECSTASY (2008)

Certain amphetamine derivatives including MDMA (ecstasy) were added to the Misuse of Drugs Act in 1977 as part of a wider generic control of ring-substituted phenethylamines. At the time, the inclusion of these substituted phenethylamines in Class A was a not unreasonable precautionary measure since little was known about their harmful effects, and misuse, at least in the UK, was almost unknown. Although

---

[14] http://nds.coi.gov.uk/Content/Detail.asp?ReleaseID = 366759&NewsAreaID = 2

Table 11.1 Classification of cannabis[a] (1971–2008).

Statute/Enquiry/Review Body	Year	Recommendation	Government response
Misuse of Drugs Act	1971	Place in Class B	
ACMD	1979	Move to Class C	Rejected
The Independent Enquiry into the Misuse of Drugs Act	2000	Move to Class C	Rejected
ACMD	2002	Move to Class C	Accepted
Home Affairs Select Committee	2002	Move to Class C	Accepted
ACMD	2006	Retain in Class C	Accepted
ACMD	2008	Retain in Class C	Rejected – to be moved to Class B

[a] Cannabis is here used as a shorthand for herbal cannabis, cannabis resin, and from 2002, cannabinol and cannabinol derivatives

ecstasy would not become common in the UK for another ten years, and much later in some European countries, MDMA is now viewed by many as much less harmful than most of the substances in Class A. In the light of twenty years experience, and despite animal studies that showed that MDMA depleted serotonin levels and caused pruning of axons, it is probably safe to conclude that there are now fewer concerns about long-term neurotoxicity. In their 2006 report, the Select Committee on Science and Technology recommended that ACMD should review the status of ecstasy. This is now underway, but it is unlikely that a recommendation will be made on the classification of MDMA before 2009.

11.14 SYSTEMATIC REVIEW OF OTHER SUBSTANCES (2009 ONWARDS)

Based on the recommendations of the numerous reviews discussed earlier, then amongst existing controlled drugs, likely possibilities for more detailed investigation by ACMD include LSD, some or all benzodiazepines and magic mushrooms/psilocin. The primary purpose of such reviews would be to determine if the current classification of the substances examined was still appropriate.

11.15 THE "PRECAUTIONARY PRINCIPLE"

The attitude of the UK Government to the 2008 report on cannabis classification by ACMD (see above) was an application of the "precautionary principle". In other words, if there is the possibility of harm then one should not do anything that might increase that harm. Even assuming that the classification of a substance has any bearing on subsequent population harm, the decision overrides the scientific risk-assessment process. It could be argued that the proper use of the precautionary principle is in situations where little information on risk is available, and the proper course of action is to err on the side of caution. That might describe, for example, the situation with MDMA and other ring-substituted phenethylamines in 1977, when they were placed in Class A. But in the case of cannabis, and despite the large amount of published data, there is still no certain answer to basic questions such as does cannabis use lead to chronic psychosis, or is high-potency cannabis necessarily more harmful. With this continued lack of knowledge, it is not entirely surprising that political reaction might be guided by caution. It might also be said that the ACMD itself has not been fully consistent in its arguments. Thus, it invoked the precautionary principle when proposing that methylamphetamine should move to Class A (see above). One could say that if it was right to move methylamphetamine to Class A on the basis of caution, and at a time when there was hardly any misuse of that drug in the UK, then the same precautionary approach should have been used for cannabis in 2008.

An interesting variation of the precautionary principle arose during the EMCDDA risk assessment of BZP in 2007 (see Chapter 9). A persuasive argument leading to its ultimate EU-wide control was based on the fact that there was no evidence that BZP was safe.

11.16 CONCLUSIONS

The final words of James Callaghan's 1970 statement concerning changes to be made in the classification in the light of new scientific knowledge may now have a hollow ring; the passage of time has led to a great increase in scientific knowledge of drugs matched only by a corresponding ossification of their classification. The placing of substances into three classes was a fairly arbitrary process since no formal risk assessment was conducted. Yet despite this, in the intervening years, few substances have been reclassified. This inertia against change has not been helped by the absence of any procedure for monitoring the impact of modifications to the Act. Part of the problem is that the classification system is a blunt weapon, and many other factors influence drug misuse beyond the potential legal consequences for offenders. Many observers have concluded that the system is largely impervious to change and should be replaced for that reason alone, although another argument would say that the system has not needed to change because the architects of the classification scheme got it almost right all those years ago.

More seriously, the classification system has been used by different stakeholders for different purposes. The original intention was to create a scale of penalties. In principle this should have been based on the relative harmfulness of controlled drugs, but later work would show that the correlation between Class and either social or individual harm was weak. For politicians, classification has been a means of showing that they are "tough on drugs" and a convenient tool to spread messages about drug harm, some of which may not always have been based on sound scientific evidence. More cynical observers would say that re-classification has become a political football, which often comes into play shortly before elections or is used as a distraction from other events. For the police and customs, particularly after the appearance of a drug strategy in the late 1990s, it became a means of prioritising law enforcement activity against those drugs associated with the greatest social harm, *i.e.* certain Class A drugs. It is clear that members of the public have only a limited concept of which drugs are in which classes, and it has been suggested that most drug users are either unaware or are unconcerned about a drug's classification. There is little doubt that, in some quarters, the reclassification of cannabis in 2004 led to at least an initial confusion about its legal status. This was not helped by the way in which the reclassification was bundled with a realignment of the penalties associated with Class C drugs. These events together led to a common view amongst criminal lawyers that the system had degenerated into a two-Class system, *i.e.* Class A drugs and everything else. Given that heroin and crack cocaine account for almost all of the social harm associated with drug misuse, the emerging *de facto* two-Class system could even be described as crack cocaine and heroin versus the rest. As noted earlier, a two-Class system was probably the original intention in the late 1960s and was allegedly the favoured solution in the Government's aborted 2006 review. It would therefore seem probable that, by accident or design, Schedule 2 of the Misuse of Drugs Act may in time be reduced to having just two parts rather than the current three.

Table 11.2 Drugs that have been reclassified (1971–2008).

Substance	Original Class	Reclassified	Date (Modification Order)
Nicodicodine	A	B	1973 (S.I. 771)
Methaqualone	C	B	1984 (S.I. 859)
Methylamphetamine	B	A	2006 (S.I. 3331)

Table 11.3 Drugs that have been temporarily reclassified (1971–2008).

Substance	Original Class	Reclassified	Date (Modification Order)	Future Classification[a]
Cannabinol and cannabinol derivatives	A	C	2003 (S.I. 3201)	B
Cannabis and cannabis resin	B	C	2003 (S.I. 3201)	B

[a] In 2003, cannabis and cannabis resin were moved from Class B to Class C, and cannabinols and cannabinol derivatives were moved from Class A to Class C (S.I. 3201). This reclassification came into effect in January 2004. In May 2008, following the third successive review of cannabis by ACMD since 2002, it was announced that cannabis, cannabis resin, cannabinol and cannabinol derivatives were to be moved to Class B in 2009.

It is also unfortunately true that, in the popular mind, the concept of reclassification has either meant reclassification to a lower Class or, more commonly, has been totally confused with declassification or even legalisation. Tables 11.2 and 11.3 show the few substances that have been reclassified since 1971. The only substances to have been permanently moved down in that time are nicodicodine, cannabinol and cannabinol derivatives; none of which could be described as commonly misused.

CHAPTER 12
The Future of "Substance" Legislation in the UK

12.1 INTRODUCTION

Following the various reviews of the Misuse of Drugs Act, and particularly the scrutiny of the classification system, many commentators have argued that the Act has outlived its usefulness or at least needs major revision. However, there are rather fewer ideas on how the legislation should be reshaped. The most commonly heard suggestions are that classification should be decoupled from penalties, that there should be a unified scale of harm based on scientific evidence, and that classification should be removed from political control. Some of these objectives may be easier to achieve than others.

The following sections provide some initial steps on how the scope of the Act might be broadened to become a "Misuse of Substances Act". In so doing it would consolidate legislative controls on other chemicals and harmful substances. While it is traditional to regard these other substances in isolated groups, they all overlap to a greater or lesser extent. A unified scheme of "substance" control could remove the anomalous double listing of some chemicals. To give a few examples: many controlled drugs and some drug precursors (*e.g.* ergotamine) are active pharmaceutical ingredients (APIs); nicotine is a poison, an API and a constituent of tobacco; formic acid and hydrochloric acid are poisons and precursors/reagents used in illicit drug synthesis, acetone is a drug and an explosives precursor, the broad group of "organophosphates"

Forensic Chemistry of Substance Misuse: A Guide to Drug Control
By L.A. King
© L.A. King 2009
Published by the Royal Society of Chemistry, www.rsc.org

are found as poisons, dangerous substances and chemical weapons; khat and many other plants contain controlled drugs, but are themselves not controlled; and cutting agents (drug diluents/adulterants) are often APIs, but are also used in the manufacture of illicit drugs.

12.2 OFFENCE-DEPENDENT CLASSIFICATION

The introduction of a scale of drug harm, as discussed in Chapter 11, would help to resolve some of the current anomalies with existing classifications. But more fundamental reform is needed if other substances are to be brought within a unified system. One of the characteristics of these other substances, be they alcohol, poisons, solvents and so on, is that there should often be no possession offence. On the other hand, most of those substances are, and should continue to be, controlled at the level of manufacture, sale and distribution. This aim could be achieved by having two classifications systems operating in parallel, one of which would apply only to possession and the other to the remaining offences. It could be argued that a partial two-dimensional classification system is already in place. Thus, there is no possession offence for substances in Part II of Schedule 4 of the Misuse of Drugs Regulations (*i.e.* anabolic steroids and related compounds) when in the form of a medicinal product. Tables 12.1 and 12.2 give an outline of how an offence-dependent classification approach might operate. This is followed by a definition of the various parts of the two-tier approach.

12.2.1 The Division of Offences into Two Groups

It is proposed that all harmful substances should be classified in two ways. Firstly, in Category I, which is used for the offences of possession or social supply and secondly in Category II, which applies to all other offences. There are many chemicals that the general public should continue to have a right to possess, but where other activities need controlling. The concept of "social supply" is not a new idea; it has previously been floated in discussions about the status of the more widely consumed controlled drugs such as ecstasy and cannabis. Many commentators have wished to see a distinction in law between the person who gives a small amount of drug to a friend and the dealer who trades in large amounts for profit. At present, although the Courts will take recognition of the particular circumstances of every case, the Misuse of Drugs Act makes no distinction. The introduction of the term "small

Table 12.1 Classification of substances in a proposed "Misuse of Substances Act": Category I (Offences of possession and social supply).

		Class		
A	B		C	D

←----------Existing controlled drugs (with some exceptions)----------→
←----------Drug precursors (UN1988 Table I)----------→ ←----------Drug precursors (UN1988 Table II)----------→
←----------Chemical weapons and precursors and explosives----------→ ←----------Poisons and dangerous chemicals----------→

Generically defined substances

Alcohol, tobacco, khat, APIs not otherwise controlled, cutting agents, volatile solvents, anabolic steroids, poppy-straw, *etc.*

Table 12.2 Classification of substances in a proposed "Misuse of Substances Act": Category II (Offences other than possession or social supply).

	Class		
A	B	C	D
------Existing controlled drugs (with some exceptions)------→			
		Generically defined substances ------→	
			Alcohol, tobacco, khat, drug precursors (UN1988 Table II), APIs not otherwise controlled, cutting agents, volatile solvents, anabolic steroids, poppy-straw, *etc.*
←------Poisons, dangerous chemicals, explosives, chemical weapons and precursors------			
	←------Drug precursors (UN1988 Table I)------→		

amount" into legislation is again not a revolutionary step; the Government intended to do just that in Section 2 of the Drugs Act 2005. Furthermore, many countries in the European Union include definitions of "small amounts" as a means of distinguishing possession from supply[1,2]. For cannabis, these amounts are typically less than 20 g, and for heroin and cocaine less than 1 g.

12.2.2 The Proposed Class D

The Class system is extended to include Class D. This has several functions, some of which are described below. In essence, Class D for the proposed Category II control includes a penalty structure that is based largely on existing controls. But Class D would also offer a means of controlling manufacture, large-scale distribution, importation and exportation for some substances such as khat that are currently unregulated. But Class D substances in the proposed Category I would have no associated possession offence. In the Misuse of Drugs Regulations, Part II of Schedule 4 would be redundant, as would the exception of poppy-straw from a possession offence (Regulation 4). Other substances that could be brought within the legislation for the first time include cutting agents, active pharmaceutical ingredients and volatile solvents, although probably only when aggravating circumstances are present. The classification illustrated here, *i.e.* A,B,C and D, is based on a simple extension of the three Classes in the current Misuse of Drugs Act. However, even if Classes B and C were to be formally collapsed into a single group, as discussed in the previous chapter, the comprehensive Act envisaged here could operate equally well as a three-Class system, *i.e.* A, B + C and D.

12.2.3 The Overall Scale of Harm

The general principle underlying the classification system is that it should be based as far as possible on an objective measure of harm. However, the concept of harm would have to be broadened from that described in Chapter 11. For example, it would need to encapsulate the potential harm caused by the unlicensed manufacture and distribution of precursor chemicals even when the intrinsic harm (*i.e.* toxic effects) of such

---

[1] *The role of the quantity in the prosecution of drug offences*, ELDD Comparative study, EMCDDA, 2003; http://eldd.emcdda.europa.eu/html.cfm/index44883EN.html

[2] Illicit drug use in the EU: legislative approaches, EMCDDA, 2005; http://www.emcdda.europa.eu/index.cfm?nNodeID = 7079

substances may be quite modest. The scale of harm could not be based solely on measurable quantities, but would also reflect existing social, political and economic factors. The example of alcohol shows that, although it may be as harmful, in a restricted definition of harm, as some Class A drugs, no social purpose would be served by making alcohol a Class A drug.

12.2.4 Generically Defined Substances

Generic definitions blur the link between harm and classification. It is certain that amongst the group of generically controlled substances are many that will have no pharmacological activity and will be essentially harmless. Yet many of those substances will sit alongside far more harmful members in the same classification group. A solution to this dilemma is to adopt the approach used in New Zealand whereby all generically controlled substances fall into the lowest classification group until otherwise shown that they are more harmful. At that point they would be removed from generic control and listed specifically in the appropriate Class. While this might sound burdensome, the reality is that of the most widely abused substances, very few are currently captured by generic controls. The only exception is MDMA, and one outcome of the current review (Chapter 11) is for it to be reclassified and listed specifically in Class B. This would mean removing MDMA from the definition of a substituted phenethylamine (Chapter 6).

12.2.5 Precursor Chemicals

The idea that precursor chemicals might be incorporated into drug legislation is not new. As noted in Appendix 4, there are several substances listed in the UN 1961 Convention that qualify as precursors, even though that Convention prefers to use the term "intermediates". In 1980, the US government added 1-phenyl-2-propanone (listed as phenylacetone) to Schedule II of the Controlled Substances Act. This precursor is used for the illicit manufacture of amphetamine and methylamphetamine (Appendix 5). There are now a number of precursors of other chemicals listed in the US legislation including: 1-phenylcyclo-hexylamine and 1-piperidinocyclohexanecarbonitrile (both Schedule II and both precursors to phencyclidine [PCP]); and lysergic acid and lysergic acid amide (both Schedule III and both precursors to lysergide [LSD]).

In Europe, a number of countries have absorbed control of precursor chemicals into their main drugs legislation. In some cases, this is an administrative convenience, while in others it means that the penalties for offences involving precursors are aligned with those for certain drugs. Insofar as the principal offences relating to precursors (*i.e.* production, supply, importation and exportation) overlap with offences relating to controlled drugs, the integration of both groups is relatively straightforward. In the US model, possession of certain precursors, like possession of other drugs, also becomes an offence. This also need cause few difficulties since some of the chemicals in question, particularly those in the proposed Category 1 (*e.g.* P2P, PMK) have few if any legitimate uses.

12.2.6 Existing Controlled Drugs

No immediate change need be made to any of the existing classifications of controlled drugs. Anabolic steroids and poppy-straw would move from Class C to Class D in both Categories. However, the introduction of a two-tier classification scheme means that existing controlled drugs need not be in the same Class for both Categories. An example here is the recently controlled ketamine where the primary objective of control was to prevent importation rather than criminalise users. Thus, ketamine could be Class C or higher in Category II, but Class D in Category I.

12.2.7 Other Chemicals

In Tables 12.1 and 12.2, the classification of poisons, dangerous chemicals, explosives, chemical weapons and their precursors is left flexible. Some risk assessment may be needed on each, but the classification system that already exists within the current legislation for these substances (Chapter 2, Appendices 7 and 8) should provide a guide.

General Bibliography

Items are listed in reverse date order:

L.A. King, S.D. McDermott, S. Jickell and A. Negrusz, *Drugs of Abuse* In: *Clarke's Analytical Toxicology*, ed. S Jickell and A. Negrusz, Pharmaceutical Press, London, 2008

Advisory Council on the Misuse of Drugs, *Classification of cannabis and public health*, Home Office, London, 2008; http://drugs.homeoffice.gov.uk/publication-search/cannabis/acmd-cannabis-report-2008

R. Murphy and S. Roe, *Drug Misuse Declared: Findings from the 2006/07 British Crime Survey*, England and Wales, Home Office, London, 2007; www.homeoffice.gov.uk/rds/pdfs07/hosb1807.pdf

EMCDDA, *Annual Report 2007: The State of the Drugs Problem in Europe*, Luxembourg, 2007; http://www.emcdda.europa.eu/html.cfm/index419EN.html

L.A. King and R Sedefov, *Early Warning System on New Psychoactive Substances: Operating Guidelines*, EMCDDA, 2007; http://www.emcdda.europa.eu/index.cfm?fuseaction=public.Content&nnodeid=431&sLanguageiso=EN

World Drug Report 2007, United Nations Office on Drugs and Crime, Vienna, 2007; http://www.unodc.org/india/world_drug_report_2007.html

E. Reid, Seizures of Drugs in England and Wales 2005, Home Office Research, Development and Statistics Directorate, London, 2007; http://www.homeoffice.gov.uk/rds/pdfs07/hosb1707.pdf

Department of Health, *United Kingdom Drug Situation, 2007 Edition, UK Focal Point on Drugs, Annual Report to the European Monitoring*

Centre for Drugs and Drug Addiction (EMCDDA), London, 2007; http://www.ukfocalpoint.org.uk/web/Publications201.asp

Drugs: Facing Facts, The Report of the RSA Commission on Illegal Drugs, Communities and Public Policy, London, 2007; http://www.rsadrugscommission.org/

L. Jason-Lloyd, *Drugs, Addiction and the Law,* 11th edn, Elm Publications, Huntingdon, 2006; http://www.elm-training.co.uk/dal11.htm #new_format

L. Iversen, *Speed, Ecstasy, Ritalin: The Science of Amphetamines*, Oxford University Press, Oxford, 2006

Advisory Council on the Misuse of Drugs, *Further consideration of the classification of cannabis under the Misuse of Drugs Act 1971*, Home Office, London, 2006; http://drugs.homeoffice.gov.uk/publication-search/acmd/cannabis-reclass-2005?view=Binary

Recommended Methods for the Identification and Analysis of Amphetamine, Methamphetamine and their Ring-Substituted Analogues in Seized Materials (revised and updated), Manual for Use by National Drug Testing Laboratories, United Nations, New York, 2006; http://www.unodc.org/unodc/en/scientists/recommended-methods-for-the-identification-.html

R. Fortson, *Misuse of Drugs: Offences, Confiscation and Money Laundering*, 5th edn, Sweet and Maxwell, London, 2005

L.A. King and S. McDermott, *Drugs of Abuse* In: *Clarke's Analysis of Drugs and Poisons*, Third edn, Vol. 1, 37–52, ed. A.C. Moffat, M.D. Osselton and B. Widdop, Pharmaceutical Press, London, 2004

Clarke's Analysis of Drugs and Poisons, 3rd edn, Vol. 2, ed. A.C. Moffat, M.D. Osselton and B. Widdop, Pharmaceutical Press, London, 2004

L.A. King, *The Misuse of Drugs Act: A Guide for Forensic Scientists*, Royal Society of Chemistry, London, 2003; http://www.rsc.org/shop/books/2003/9780854046256.asp

M.D. Cole, *The Analysis of Controlled Substances: A Systematic Approach*, Wiley, New York, 2003

Ecstasy and Amphetamines Global Survey 2003, United Nations Office on Drugs and Crime, Vienna and New York, 2003; http://www.unodc.org/pdf/publications/report_ats_2003-09-23_1.pdf

House of Commons Home Affairs Committee, *The Government's Drugs Policy: Is it Working?*, Third Report of Session 2001–02, Volume 1: Report and Proceedings of the Committee, The Stationery Office, London, 2002; http://www.parliament.the-stationery-office.com/pa/cm200102/cmselect/cmhaff/318/31804.htm

Advisory Council on the Misuse of Drugs, *The Classification of Cannabis under the Misuse of Drugs Act 1971*, Home Office, London, 2002;

http://drugs.homeoffice.gov.uk/publication-search/acmd/cannabis-class-misuse-drugs-act?view=Binary

The Police Foundation, *Drugs and the Law: Report of the Independent Inquiry into the Misuse of Drugs Act 1971*, London, 2000

United Nations Office on Drugs and Crime, *Multilingual Dictionary of Narcotic Drugs and Psychotropic Substances under International Control*, United Nations, New York, 2006; http://www.unodc.org/unodc/en/press/releases/2007-02-05.html

A. Shulgin and A. Shulgin, *TIHKAL, The Continuation*, Transform Press, Berkeley, California, 1997

Parliamentary Office of Science and Technology, *Common Illegal Drugs and their Effects*, House of Commons, London, 1996

A. Shulgin and A. Shulgin, *PIHKAL: A Chemical Love Story*, Transform Press, Berkeley, California, 1991

M. Klein, F. Sapienza, H. McClain and I. Khan, I., (ed.), *Clandestinely Produced Drugs, Analogues and Precursors: Problems and Solutions*, United States Department of Justice Drug Enforcement Administration, Washington, D.C., 1989

APPENDIX 1
Modification and Amendment Orders to the Misuse of Drugs Act 1971

A1.1 THE MISUSE OF DRUGS ACT 1971 (MODIFICATION) ORDER 1973 (S.I. 771)

Transfers nicodicodine from Part I to Part II of Schedule 2 and excludes from Part I certain substances (notably, codeine, dihydrocodeine, ethylmorphine, norcodeine and pholcodine) that are already included in Part II. The Order also adds drotebanol to Part I and propiram to Part II and removes fencamfamin, pemoline, phentermine and prolintane from Part III.

A1.2 THE MISUSE OF DRUGS ACT 1971 (MODIFICATION) ORDER 1975 (S.I. 421)

Adds difenoxin and 4-bromo-2,5-dimethoxy-α-methylphenethylamine to Part I of Schedule 2.

A1.3 THE MISUSE OF DRUGS ACT 1971 (MODIFICATION) ORDER 1977 (S.I. 1243)

Adds certain tryptamine derivatives and certain phenethylamine derivatives to Part I of Schedule 2.

A1.4 THE MISUSE OF DRUGS ACT 1971 (MODIFICATION) ORDER 1979 (S.I. 299)

Adds phencyclidine to Part I of Schedule 2.

A1.5 THE MISUSE OF DRUGS ACT 1971 (MODIFICATION) ORDER 1983 (S.I. 765)

Adds sufentanil and tilidate to Part I of Schedule 2 and dextropropoxyphene to Part III.

A1.6 THE MISUSE OF DRUGS ACT 1971 (MODIFICATION) ORDER 1984 (S.I. 859)

Adds alfentanil, eticyclidine, rolicyclidine and tenocyclidine to Part I of Schedule 2, certain barbiturates (that is to say 5,5-disubstituted barbituric acids and methylphenobarbitone) and mecloqualone to Part II, transfers methaqualone from Part III to Part II and adds diethylpropion to Part III.

A1.7 THE MISUSE OF DRUGS ACT 1971 (MODIFICATION) ORDER 1985 (S.I. 1995)

Adds glutethimide, lefetamine and pentazocine to Part II of Schedule 2, removes explicit reference to dexamphetamine, and adds ethchlorvynol, ethinamate, mazindol, meprobamate, methyprylone, phentermine and a group of 33 benzodiazepines to Part III.

A1.8 THE MISUSE OF DRUGS ACT 1971 (MODIFICATION) ORDER 1986 (S.I. 2230)

Adds carfentanil, lofentanil, certain fentanyl derivatives and certain pethidine derivatives to Part I of Schedule 2 and cathine, cathinone, fencamfamin, fenethylline, fenproporex, mefenorex, propylhexedrine, pyrovalerone and N-ethylamphetamine to Part III.

A1.9 THE MISUSE OF DRUGS ACT 1971 (MODIFICATION) ORDER 1989 (S.I. 1340)

Adds buprenorphine and pemoline to Part III of Schedule 2.

A1.10 THE MISUSE OF DRUGS ACT 1971 (MODIFICATION) ORDER 1990 (S.I. 2589)

Adds N-hydroxy-tenamphetamine and 4-methyl-aminorex to Part I of Schedule 2 and midazolam to Part III.

A1.11 THE MISUSE OF DRUGS ACT 1971 (MODIFICATION) ORDER 1995 (S.I. 1966)

Removes propylhexedrine from Part III of Schedule 2.

A1.12 THE MISUSE OF DRUGS ACT 1971 (MODIFICATION) ORDER 1996 (S.I. 1300)

Adds certain anabolic/androgenic steroids, clenbuterol and certain polypeptide hormones to Part III of Schedule 2.

A1.13 THE MISUSE OF DRUGS ACT 1971 (MODIFICATION) ORDER 1998 (S.I. 750)

Adds etryptamine to Part I of Schedule 2, methcathinone and zipeprol to Part II and aminorex, brotizolam and mesocarb to Part III.

A1.14 THE MISUSE OF DRUGS ACT 1971 (MODIFICATION) ORDER 2001 (S.I. 3932)

Adds 35 phenethylamine derivatives to Part I of Schedule 2 and α-methylphenethylhydroxylamine to Part II.

A1.15 THE MISUSE OF DRUGS ACT 1971 (MODIFICATION) ORDER 2003 (S.I. 1243)

Adds dihydroetorphine and remifentanil to Part I of Schedule 2, and 4-hydroxy-n-butyric acid, zolpidem, 4-androstene-3,17-dione, 5-androstene-3,17-diol, 19-nor-4-androstene-3,17-dione and 19-nor-5-androstene-3,17-diol to Part III.

A1.16 THE MISUSE OF DRUGS ACT 1971 (MODIFICATION) (NO. 2) ORDER 2003 (S.I. 3201)

Moves cannabinol and cannabinol derivatives from Part I of Schedule 2 to Part III and moves cannabis and cannabis resin from Part II of Schedule 2 to Part III.

A1.17 THE MISUSE OF DRUGS ACT 1971 (AMENDMENT) ORDER 2005 (S.I. 3178)

Adds ketamine to Part III of Schedule 2.

A1.18 THE MISUSE OF DRUGS ACT 1971 (AMENDMENT) ORDER 2006 (S.I. 3331)

Moves methylamphetamine from Part II of Schedule 2 to Part I.

APPENDIX 2
The Misuse of Drugs Regulations (Schedule 4)

Schedule 4 of the Regulations has two parts; the full text of each is shown below. This is based on the original text of the Misuse of Drugs Regulations 2001 as amended. Thus, the Misuse of Drugs (Amendment) Regulations 2003 (S.I. 1432)[1] added zolpidem and 4-hydroxy-n-butyric acid to Part I and added 4-androstene-3,17-dione, 5-androstene-3,17-diol, 19-nor-4-androstene-3,17-dione and 19-nor-5-androstene-3,17-diol to Part II. The Misuse of Drugs (Amendment) (No. 3) Regulations 2005 (S.I. 3372)[2] added ketamine to Part I. The Misuse of Drugs and Misuse of Drugs (Safe Custody) (Amendment) Regulations 2007 (S.I. 2154)[3] moved midazolam from Schedule 4 Part I to Schedule 3 with effect from 1 January 2008.

---

[1] http://www.opsi.gov.uk/si/si2003/20031432.htm
[2] http://www.opsi.gov.uk/si/si2005/20053372.htm
[3] http://www.opsi.gov.uk/si/si2007/uksi_20072154_en_1

Forensic Chemistry of Substance Misuse: A Guide to Drug Control
By L.A. King
© L.A. King 2009
Published by the Royal Society of Chemistry, www.rsc.org

Table A2.1 Full text of Schedule 4 Part I as it appears in the Regulations Controlled Drugs Subject to the Requirements of Regulations 22, 23, 26 and 27.

1. Alprazolam
   Aminorex
   Bromazepam
   Brotizolam
   Camazepam
   Chlordiazepoxide
   Clobazam
   Clonazepam
   Clorazepic acid
   Clotiazepam
   Cloxazolam
   Delorazepam
   Diazepam
   Estazolam
   Ethyl loflazepate
   Fencamfamin
   Fenproporex
   Fludiazepam
   Flurazepam
   Halazepam
   Haloxazolam
   4-Hydroxy-n-butyric acid
   Ketamine
   Ketazolam
   Loprazolam
   Lorazepam
   Lormetazepam
   Medazepam
   Mefenorex
   Mesocarb
   Nimetazepam
   Nitrazepam
   Nordazepam
   Oxazepam
   Oxazolam
   Pemoline
   Pinazepam
   Prazepam
   Pyrovalerone
   Tetrazepam
   Triazolam
   N-Ethylamphetamine
   Zolpidem
2. Any stereoisomeric form of a substance specified in paragraph 1.
3. Any salt of a substance specified in paragraph 1 or 2.
4. Any preparation or other product containing a substance or product specified in any of paragraphs 1 to 3, not being a preparation specified in Schedule 5.

Table A2.2 Full text of Schedule 4 Part II as it appears in the Regulations.

Controlled drugs excepted from the prohibition on possession when in the form of a medicinal product; excluded from the application of offences arising from the prohibition on importation and exportation when imported or exported in the form of a medicinal product by any person for administration to himself; and subject to the requirements of Regulations 22, 23, 26 and 27.

1. The following substances, namely
   4-Androstene-3,17-dione
   5-Androstene-3,17-diol
   Atamestane
   Bolandiol
   Bolasterone
   Bolazine
   Boldenone
   Bolenol
   Bolmantalate
   Calusterone
   Methenolone
   Methyltestosterone
   Metribolone
   Mibolerone
   Nandrolone
   19-Nor-4-androstene-3,17-dione
   19-Nor-5-androstene-3,17-diol
   Norboletone
   Norclostebol
   Norethandrolone

Table A2.2 (*Continued*).

4-Chloromethandienone	Ovandrotone
Clostebol	Oxabolone
Drostanolone	Oxandrolone
Enestebol	Oxymesterone
Epitiostanol	Oxymetholone
Ethyloestrenol	Prasterone
Fluoxymesterone	Propetandrol
Formebolone	Quinbolone
Furazabol	Roxibolone
Mebolazine	Silandrone
Mepitiostane	Stanolone
Mesabolone	Stanozolol
Mestanolone	Stenbolone
Mesterolone	Testosterone
Methandienone	Thiomesterone
Methandriol	Trenbolone

2. Any compound (not being Trilostane or a compound for the time being specified in paragraph 1 of this Part of this Schedule structurally derived from 17-hydroxyandrostan-3-one or from 17-hydroxyestran-3-one by modification in any of the following ways, that is to say
    (i) by further substitution at position 17 by a methyl or ethyl group;
    (ii) by substitution to any extent at one or more of the positions 1,2,4,6,7,9,11 or 16, but at no other position;
    (iii) by unsaturation in the carbocyclic ring system to any extent, provided that there are no more than two ethylenic bonds in any one carbocyclic ring;
    (iv) by fusion of ring A with a heterocyclic system.
3. Any substance that is an ester or ether (or, where more than one hydroxyl function is available, both an ester and an ether) of a substance specified in paragraph 1 or described in subparagraph 2 of this Part of this Schedule.
4. The following substances, namely
    Chorionic gonadotrophin (HCG)
    Clenbuterol
    Nonhuman chorionic gonadotrophin
    Somatotropin
    Somatrem
    Somatropin
5. Any stereoisomeric form of a substance for the time being specified or described in any of the paragraphs 1 to 4 of this Part of this Schedule.
6. Any salt of a substance specified or described in any of paragraphs 1 to 6 of this Part of this Schedule.
7. Any preparation or other product containing a substance or product specified or described in any of paragraphs 1 to 6 of this Part of this Schedule, not being a preparation specified in Schedule 5.

APPENDIX 3
The Misuse of Drugs Regulations (Schedule 5)

The following is the full text of Schedule 5 as it appears in the Misuse of Drugs Regulations 2001 as amended by the Misuse of Drugs (Supply to Addicts) (Amendment) Regulations 2005 (S.I. 2864)[1]. Regulation 13 revoked the original paragraph 2 of Schedule 5. This formerly read: *"Any preparation of cocaine containing not more than 0.1% of cocaine calculated as cocaine base, being a preparation compounded with one or more other active or inert ingredients in such a way that the cocaine cannot be recovered by readily applicable means or in a yield which would constitute a risk to health."* This original text had its origins in UN1961 and had been included to ease the regulatory burden on those using cocaine for therapeutic use. However, by the 1990s it was clear that not only was therapeutic use of cocaine limited, but physicians and hospital staff usually kept cocaine in more concentrated solutions to which the Regulations in Schedule 3 applied. More significantly, the original Regulation had become a burden to forensic science laboratories, which needed to determine the concentration of cocaine in every seizure. Since almost all illicit cocaine has a purity far in excess of 0.1%, the cost of such analysis became difficult to justify.

Controlled Drugs Excepted from the Prohibition on Importation, Exportation and Possession and Subject to the Requirements of Regulations 24 and 26.

---

[1] http://www.opsi.gov.uk/si/si2005/20052864.htm

Forensic Chemistry of Substance Misuse: A Guide to Drug Control
By L.A. King
© L.A. King 2009
Published by the Royal Society of Chemistry, www.rsc.org

The Misuse of Drugs Regulations (Schedule 5) 159

1. (1) Any preparation of one or more of the substances to which this paragraph applies, not being a preparation designed for administration by injection, when compounded with one or more other active or inert ingredients and containing a total of not more than 100 milligrams of the substance or substances (calculated as base) per dosage unit or with a total concentration of not more than 2.5% (calculated as base) in undivided preparations.
   (2) The substances to which this paragraph applies are acetyldihydrocodeine, codeine, dihydrocodeine, ethylmorphine, nicocodine, nicodicodine (6-nicotinoyldihydrocodeine), norcodeine, pholcodine and their respective salts.
2. Any preparation of medicinal opium or of morphine containing (in either case) not more than 0.2% of morphine calculated as anhydrous morphine base, being a preparation compounded with one or more other active or inert ingredients in such a way that the opium or, as the case may be, the morphine cannot be recovered by readily applicable means or in a yield which would constitute a risk to health.
3. Any preparation of dextropropoxyphene, being a preparation designed for oral administration, containing not more than 135 milligrams of dextropropoxyphene (calculated as base) per dosage unit or with a total concentration of not more than 2.5% (calculated as base) in undivided preparations.
4. Any preparation of difenoxin containing, per dosage unit, not more than 0.5 milligrams of difenoxin and a quantity of atropine sulphate equivalent to at least 5% of the dose of difenoxin.
5. Any preparation of diphenoxylate containing, per dosage unit, not more than 2.5 milligrams of diphenoxylate calculated as base, and a quantity of atropine sulphate equivalent to at least 1% of the dose of diphenoxylate.
6. Any preparation of propiram containing, per dosage unit, not more than 100 milligrams of propiram calculated as base and compounded with at least the same amount (by weight) of methylcellulose.
7. Any powder of ipecacuanha and opium comprising -
   10% opium, in powder,
   10% ipecacuanha root, in powder, well mixed with 80% of any other powdered ingredient containing no controlled drug.
8. Any mixture containing one or more of the preparations specified in paragraphs 1 to 8, being a mixture of which none of the other ingredients is a controlled drug.

APPENDIX 4
Drug "Intermediates" in the Misuse of Drugs Act 1971

Not all substances listed in the Act are abusable as such; there are several examples of drug precursors/intermediates. They are all in Class A and all derive from UN1961 (see Table A4.1). Although γ-butyrolactone (GBL) is a synthetic precursor to γ-hydroxybutyrate (GHB), ACMD recommended in 2008 that it should be added to the Misuse of Drugs Act. It is not included here since it is also a metabolic precursor to GHB, and therefore best regarded as an active substance.

Table A4.1 Drug "intermediates" listed in the Act.

Substance	Alternative name
4-Cyano-2-dimethylamino-4, 4-diphenylbutane	Methadone Intermediate
4-Cyano-1-methyl-4-phenylpiperidine	Pethidine Intermediate A
Ecgonine	Cocaine precursor
2-Methyl-3-morpholino-1,1-diphenylpropane-carboxylic acid	Moramide Intermediate
1-Methyl-4-phenylpiperidine-4-carboxylic acid	Pethidine Intermediate C
4-Phenylpiperidine-4-carboxylic acid ethyl ester	Pethidine Intermediate B

Forensic Chemistry of Substance Misuse: A Guide to Drug Control
By L.A. King
© L.A. King 2009
Published by the Royal Society of Chemistry, www.rsc.org

APPENDIX 5
Drug Precursors

Tables A5.1 to A5.3 show the 23 precursors and essential reagents listed in UN1988 and The Criminal Justice (International Cooperation Act) 1990 as modified[1]. Substances in Table I of UN1988 are mostly primary precursors; that is to say they may be converted into the target drug in a single stage, whereas substances in Table II of UN1988 are either secondary precursors, where two or more chemical reactions are needed for conversion

Table A5.1 Category 1 chemicals[a].

Substance	UN1988	Typical drug products
N-Acetylanthranilic acid	}	Methaqualone
Ephedrine	}	Methylamphetamine
Ergometrine[b]	}	Lysergide (LSD)
Ergotamine	}	Lysergide (LSD)
Isosafrole	} Table 1	MDMA, MDA, *etc.*
Lysergic acid	}	Lysergide (LSD)
3,4-Methylenedioxyphenylpropan-2-one	}	MDMA, MDA, *etc.*
Norephedrine	}	Amphetamine
1-Phenyl-2-propanone	}	Amphetamine
Piperonal	}	MDMA, MDA, *etc.*
Pseudoephedrine	}	Methylamphetamine
Safrole	}	MDMA, MDA, *etc.*

[a]The legislation adds that the salts of the substances listed in Category 1 are also subsumed whenever the existence of such salts is possible
[b]Ergometrine is also known as ergonovine

---

[1] The Controlled Drugs (Drug Precursors) (Intra-Community Trade) Regulations 2008 (S.I. 2008/295) and The Controlled Drugs (Drug Precursors) (Community External Trade) Regulations 2008 (S.I. 2008/296). See also: http://drugs.homeoffice.gov.uk/drugs-laws/licensing/precursor-forms/

Forensic Chemistry of Substance Misuse: A Guide to Drug Control
By L.A. King
© L.A. King 2009
Published by the Royal Society of Chemistry, www.rsc.org

Table A5.2 Category 2 chemicals.

Substance	UN1988	Typical drug products
Acetic anhydride	} Table 1	Heroin
Potassium permanganate		Cocaine processing
Anthranilic acid	} Table II	Methaqualone
Phenylacetic acid		Amphetamine
Piperidine		Phencyclidine

Table A5.3 Category 3 chemicals.

Substance	UN1988	Typical drug products
Acetone		
Ethyl ether		
Methyl ethyl ketone		} Mostly used for
Toluene	} Table II	cocaine processing
Sulphuric acid		
Hydrochloric acid		

(*e.g.* anthranilic acid, phenylacetic acid, piperidine) or they are essential reagents and solvents. To a large extent, the solvents (all in Category 3) are used in cocaine processing. The tables also show the drug most commonly produced from illicit use of these precursors. Beyond these statutory controls, the Chemical Industries Association in the UK operates a voluntary list of additional chemicals. A similar scheme has been organised by the European Commission, while the UN gives guidance through an even more comprehensive list of drug-related chemical precursors and essential reagents (the "Limited International Special Surveillance List").

These additional chemicals include, for example: allyl benzene, ammonium formate, benzaldehyde, benzyl chloride, benzyl cyanide, cyclohexanone, diethylamine, ethylidine diacetate, ethylamine, formamide, hydriodic acid, hydrogen chloride gas, iodine, isatoic anhydride, methylamine, N-methylformamide, 2-methyl-2-(phenylmethyl)-1,3-dioxolan, 3-methyl-2-phenyloxirane, N-methylephedrine, nitroethane, phenyl-2-propanol, piperonyl alcohol, propionic anhydride, propylbenzene, o-toluidine, and red phosphorus. The substances 2-methyl-2-(phenylmethyl)-1,3-dioxolan and 3-methyl-2-phenyloxirane are effectively "pro-precursors"; both are readily convertible to 1-phenyl-2-propanone. Some of the above substances are controlled in the US. In mid-2008, the Drug Enforcement Administration added N-phenethyl-4-piperidone (a fentanyl precursor) to the US legislation[2].

---

[2] http://www.deadiversion.usdoj.gov/fed_regs/rules/2008/fr0725.htm

Structure (A5.1) shows the relationship between amphetamine and one of its listed precursors, 1-phenyl-2-propanone (P2P, also known as benzylmethylketone, phenylacetone, BMK). Structure (A5.2) shows the relationship between MDA and one of its listed precursors, 3,4-methylenedioxy-phenylpropan-2-one (also known as piperonylmethylketone, PMK).

As examples of how much drug can be manufactured from precursor chemicals, it is estimated that 20 L of P2P would produce 10 kg amphetamine, sufficient for around 400 000 doses, and 20 L of PMK would produce 200 000 ecstasy tablets. In the manufacture of heroin, 250 L of acetic anhydride is required to produce 100 kg heroin, sufficient for around 1 million doses.

Structure (A5.1) The conversion of P2P (1) to amphetamine (2)

Structure (A5.2) The conversion of PMK (1) to MDA (2)

APPENDIX 6
A Brief History of the Legal Status of Hash Oil

Preparations or products of Class B drugs are also controlled by virtue of paragraph 4 of Part II of Schedule 2, namely "*Any preparation or other product containing a substance or product for the time being specified in any of paragraphs 1 to 3 of this Part of this Schedule, not being a preparation falling within paragraph 6 of Part I of this Schedule*". Hash oil (cannabis oil) is not considered to fall within this definition because it cannot be said to contain cannabis or cannabis resin as such. However, Section 37 of the Act defines cannabis resin as " . . . *the separated resin, whether crude or purified, obtained from any plant of the genus Cannabis*". It had been accepted that hash oil is a purified form of cannabis resin and therefore (before reclassification in 2004), a Class B drug.

This situation was brought into confusion after 1990 when it became clear that some hash oil was being produced, not from cannabis resin, but from herbal cannabis. In R-v-Carter[1] it was successfully agued that such hash oil could no longer be deemed to be a purified form of cannabis resin; the only option open to the Court was to regard it as a preparation of cannabinol and therefore a Class A drug.

It is sometimes possible to distinguish the two types of hash oil insofar as cannabidiol (CBD) is present in much greater amounts in cannabis resin than it is in herbal cannabis[2]. Thus, if CBD is found in hash oil,

---

[1] Oxford Crown Court, Judgement of 16 December 1992
[2] S. Hardwick and L. King, *Home Office Cannabis Potency Study 2008*, Home Office Scientific Development Branch, St. Albans, 2008 http://drugs.homeoffice.gov.uk/publication-search/cannabis/potency

then it has probably not originated from herbal cannabis. Furthermore, hash oil made from herbal cannabis may also appear to have a dark green colour because of the presence of the pigment chlorophyll. But from the user's viewpoint, there is little distinction between the two types of hash oil; it is anomalous that it could be treated as either Class A or Class B solely on the basis of its manufacturing route.

Following the above judgement, there arose a general scientific agreement that a solution to the problem would be to define hash oil as a Class B drug by including it as a named substance in Part II of Schedule 2 to the Act. Unfortunately, unlike chemically defined substances, hash oil as an entity cannot be added to the Act by a simple Modification Order. The reason for this is that firstly a definition of hash oil would be needed in Part IV of Schedule 2 to the Act if not also in Section 37. Secondly, the definition of the Class A "cannabinols" in Part IV of Schedule 2 would have to be reworded to say "*cannabinol derivatives means the following substances, except where contained in cannabis, cannabis resin or liquid cannabis . . . .*". These changes would have required primary legislation, but a suitable definition might have been "Liquid cannabis is a solvent extract of cannabis or cannabis resin". It is interesting to note that UN1961 included the concept of liquid cannabis, but defined it as " . . . *extracts and tinctures of cannabis*". This was intended as a reference to the medicinal products once found in Pharmacopoeias, but now long obsolete.

The original separation in the Act between the Class A "cannabinols" and the corresponding Class B plant material was the only instance where the classification of a pharmacologically active substance (*i.e.* THC) was effectively based on the potency of different products or preparations. Hash oil formed an awkward bridge between these two groups. These problems were resolved in 2004 by placing cannabis and "cannabinols" into Class C (and Class B from 2009).

APPENDIX 7
Other Drug-Related Legislation

Parts of the following statutes either modify some Sections of the Misuse of Drugs Act or have some relevance to other offences involving controlled drugs. Not all necessarily refer to all countries of the UK. All UK legislation since 1988 can be found at http://www.opsi.gov.uk/legislation/about_legislation.htm. A European legal database on drugs showing country profiles is available at http://eldd.emcdda.europa.eu/.

A7.1 ROAD TRAFFIC ACT 1972

Makes it an offence to be in charge of a motor vehicle while unfit to drive through drink or drugs (controlled or otherwise).

A7.2 THE CUSTOMS AND EXCISE MANAGEMENT ACT 1979

Extends the limited powers in the Misuse of Drugs Act against importation and exportation of controlled drugs. Provides the means for prosecuting drug offenders involved in these activities.

A7.3 THE DRUG TRAFFICKING ACT 1994

Enables the UK to meet further obligations under UN1988. It replaced the Drug Trafficking Offences Act 1986. Gives police the power to seize assets and income of anyone who is found guilty of drugs trafficking, even if that income is not related to the trafficking of drugs. The Act applies to England and Wales only.

Forensic Chemistry of Substance Misuse: A Guide to Drug Control
By L.A. King
© L.A. King 2009
Published by the Royal Society of Chemistry, www.rsc.org

A7.4 THE CRIME AND DISORDER ACT 1998

Requires offenders, who are convicted of crime committed in order to fund their drug habit, to be tested for drug misuse and to undertake drug treatment. The Drug Testing and Treatment Orders (DTTOs) were later consolidated by the Powers of Criminal Courts (Sentencing) Act 2000. However, many of the provisions of the latter Act including the DTTOs are in the process of being repealed.

A7.5 THE CRIMINAL JUSTICE AND POLICE ACT 2001

Among other powers, this amends Section 8(d) of the Misuse of Drugs Act 1971 to read "administering or using a controlled drug in any person's possession at or immediately before the time when it is administered or used". Section 8(d) was previously only concerned with "smoking cannabis, cannabis resin or prepared opium". Although primarily designed to give police additional powers to close "crack houses" the proposed amendment proved controversial and has been abandoned.

A7.6 THE CRIMINAL JUSTICE ACT 2003

Among other provisions, this amends sentences for Class C drugs. Whereas the maximum penalty for possession had been five years, this was reduced to 2 years imprisonment. An amendment to the Police and Evidence Act 1984 allows the power of arrest to be used for the possession of cannabis and any other Class C drug. Before reclassification of cannabis in 2004, the possession of a Class C drug had not been an arrestable offence. The maximum prison sentence for supplying any Class C drug was increased from 5 years to 14 years, *i.e.* similar to the penalty associated with Class B drugs.

A7.7 THE CRIMINAL JUSTICE (INTERNATIONAL COOPERATION) ACT 1990; CONTROLLED DRUGS (DRUG PRECURSORS) (INTRA-COMMUNITY TRADE) REGULATIONS 2008; CONTROLLED DRUGS (DRUG PRECURSORS) (COMMUNITY EXTERNAL TRADE) REGULATIONS 2008

Part of a group of statutes created to discharge UK responsibilities to UN1988 and the corresponding EU legislation. Regulate the licensing, manufacture and distribution of substances (precursors) useful for the production of illicit drugs.

APPENDIX 8
Relevant Stated Cases

There is a substantial body of case law concerning offences under the Act. Information is given in Appendix 9 on Appeal Court hearings which led to sentencing guidelines. The following is intended as no more than a brief index to selected cases of significance to forensic scientists. It does not include those of purely historical interest such as R-v-Watts, which concerned the now obsolete inclusion of dexamphetamine in the Act, or the case of R-v-Goodchild: a legal saga that revolved around the now defunct original definition of cannabis.

A8.1 USABILITY

The currently held view is that the principle of *de minimis*[1] does not operate and that the proper approach to what was once called the "usability test" is whether the prosecution can prove knowledge of possession. In other words, apart from those clearly defined situations to which Schedule 5 of the Regulations apply, the Act defines no minimum quantities below which an offence cannot be committed. This is set out in R-v-Boyesen, 75 Cr.App.R. 51, H.L. (1982), and superseded an earlier opposing argument that had been reached in R-v-Carver, 67 Cr.App.R. 352 (1978), where it had been maintained that a defendant had to be in possession of a usable quantity of a drug.

---

[1] From the epithet "*de minimis non curat lex*": the law is not concerned with trivialities

A8.2 GENERIC LEGISLATION

In considering the generic definition of phenethylamines, the Court of Appeal in R-v-Couzens and Frankel, Cr.L.R. 822 (1992), upheld the view that in paragraph 1(c) of Part I of Schedule 2 to the Act, the term *"structurally derived from"* does not describe a process, but rather defines certain controlled drugs in terms of their molecular structure.

A8.3 CANNABIS AND CANNABIS RESIN

In R-v-Best, 70 Cr.App.R. 21 (1979), it was held that the principle of duplicity is not compromised by a prosecution for possession of cannabis or cannabis resin. In other words it is not always necessary for one or the other to be separately specified. This is likely to be of significance only in those cases where there is insufficient material for unequivocal identification of one or the other.

A8.4 CRACK COCAINE

Even though cocaine and its salts are all treated equally as Class A drugs, in R-v-Russell, 94 Cr.App.R. 351 (1992), the production of crack (*i.e.* cocaine base) from cocaine hydrochloride was deemed to be an offence of production. By extension this would seem to apply to any salt–base interconversion and does not compromise the principle that if no production is alleged then the prosecution is not required to identify the form of drug present.

A8.5 SALTS AND STEREOISOMERS

The case in R-v-Greensmith, 1 W.L.R. 1/24 (1983), was concerned with the specific example of cocaine, but the general point was established that the prosecution does not have to prove whether a controlled drug is in a particular stereoisomeric form or as a particular salt.

APPENDIX 9
Sentencing Guidelines

Maximum penalties for drugs offences are set out in the Act. Since 1982, and until recent years, sentencing in many of the larger drugs cases was based on the so-called Aramah equation. The principle was that penalties should relate to the value of the drugs seized. The street price per gram was multiplied by the weight of the seizure to get a total price. This was modified to take account of the actual purity. Thus, if the drugs were above average purity then an upward correction was made to the value. However, it had been the practice of the former HM Customs and Excise (now HM Revenue and Customs) not to decrease the value if the purity was below the average. If the value was calculated at £100 000 or more, then a sentence of 10 years imprisonment was likely. If the value was £1 million or more, then it was 14 years. As an example, 1 kg of heroin could be worth up to £100 000. Appendix 12 and Appendix 13 show typical current purities and prices, respectively, for the common illicit drugs, and which might be used in an Aramah calculation.

The Aramah equation had two major problems. Firstly, street price was subjective and often difficult to define, particularly if it was not clear on which "street" or even in which country the drugs were to be sold. In many cases, the Court had little option but to take the average value provided by the prosecution and the defence. A more fundamental objection was that if drugs become widely available, then the price would drop. The expected sentence would then also fall leading to an unacceptable situation.

The solution to these problems came about by rejecting drug valuation for the larger cases and replacing it with a more objective system based on either the weight of the pure drug or the number of dosage units. Tariffs were set by the Court of Appeal (Criminal Division) for the major Class A drugs: heroin, cocaine, the ecstasy group (MDMA *etc.*) and LSD as well as for the major Class B drugs, as shown in Tables A9.1 and A9.2. It will be noted from the tables below that there is a nonlinear relationship between sentences and weights/doses, such that smaller amounts of drug attract a disproportionately high sentence.

Because of the amounts involved, these guidelines are largely restricted to those cases involving importation, but can apply to possession with intent to supply. Although no strict ruling was given, it is generally assumed that the weights of drugs shown refer to the base rather than some (arbitrary) salt form. The guidelines were based on cases where the trial has been contested and where the defendant played no more than a subordinate role. The Court of Appeal stressed that these criteria were merely one factor in deciding appropriate sentences,

Table A9.1 Sentencing guidelines for Class A drugs.

Drug	Weight/Dose	Sentence	Case reference
Cocaine/Heroin	500 g	10 years	}R-v-Aranguren, 16 Cr.App.R. 211 (1995)
Cocaine/Heroin	5 kg	14 years	}
MDMA, *etc.*	5000 tablets	10 years	}R-v-Warren and Beeley, 1 Cr.App.R. 120 (1996)
MDMA, *etc.*	50 000 tablets	14 years	}
Lysergide (LSD)	25 000 units	10 years	}R-v-Hurley, 2 Cr.App.R. (S) 299 (1998)
Lysergide (LSD)	250 000 units	14 years	}
Opium	> 4 kg	10 years	} R-v-Mashaolli, Cr.L R. 1029 (2000)
Opium	> 40 kg	14 years	}

Table A9.2 Sentencing guidelines for Class B drugs.

Drug	Weight	Sentence	Case reference
Amphetamine	< 500 g	2 years	}R-v-Wijs *et al.*, Cr.L.R. 587 (1998)
Amphetamine	> 500 g < 2.5 kg	2–4 years	}
Amphetamine	> 2.5 kg < 10 kg	4–7 years	}
Amphetamine	> 10 kg < 15 kg	7–10 years	}
Amphetamine	> 15 kg	10–14 years	}
Cannabis	100 kg	7–8 years	}R-v-Ronchetti, Cr.L.R. 227 (1998)
Cannabis	500 kg	10 years	}

and that the role of the offender, his plea and any assistance he might have given to the authorities were examples of other considerations that the Court would have to weigh.

At the time of the judgement, MDMA dosage units typically contained 100 mg of active drug such that 5000 tablets were equal to 500 g of pure drug. However, by 2008, the typical content of ecstasy tablets had fallen to 70 mg.

Cannabis and cannabis resin should be treated equally. Unlike other drugs, where the purity or drug content needs to be taken into account, the weight of cannabis is taken to be the seizure weight, *i.e.* no notional correction to the equivalent of pure tetrahydrocannabinol is made. For cannabis oil, 1 kg should be taken as equivalent to 10 kg of cannabis or cannabis resin.

APPENDIX 10
Profiles of the Major Drugs of Misuse[1]

A10.1 AMPHETAMINE

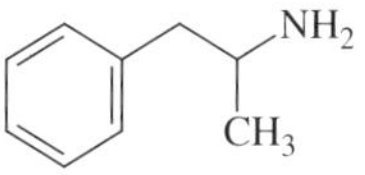

Structure (A10.1) Amphetamine

A10.1.1 Introduction

The world-wide production and consumption of amphetamine and the closely related methylamphetamine (see below) show clear geographical trends. In Europe, amphetamine is much more common than methylamphetamine, but in North America and the Far East this situation is reversed. A synthetic substance, amphetamine is normally seen as a white powder and acts as a stimulant of the central nervous system (CNS). It is believed that amphetamine was first manufactured in the 1880s by the German chemist Leuckart, although documentary evidence for this is lacking. Like methylamphetamine, it appears that systematic studies of its chemistry did not come about until the early 20th century. Amphetamine has some limited therapeutic use, but

---

[1] These profiles (apart from lysergide) are based on those originally produced by the author for the website of the EMCDDA: http://www.emcdda.europa.eu/index.cfm?nnodeid = 25328

Forensic Chemistry of Substance Misuse: A Guide to Drug Control
By L.A. King
© L.A. King 2009
Published by the Royal Society of Chemistry, www.rsc.org

most is manufactured in clandestine laboratories in Europe. It is under international control and closely related to methylamphetamine.

A10.1.2 Chemistry

Amphetamine (Chemical Abstracts System: CAS-300-62-9; Structure (A10.1)) is a member of the phenethylamine family, which includes a range of substances that may be stimulants, entactogens or hallucinogens. Thus, amphetamine is α-methylphenethylamine. The fully systematic name is α-methylbenzeneethanamine. The asymmetric α-carbon atom gives rise to two enantiomers. These two forms were previously called the (−) or l-stereoisomer and the (+) or d-stereoisomer, but in modern usage are defined as the R and S stereoisomers.

A10.1.3 Physical form

Amphetamine base is a colourless volatile oil insoluble in water. The most common salt is the sulfate (CAS-60-13-9): a white or off-white powder soluble in water. Illicit products mostly consist of powders. Tablets containing amphetamine may carry logos similar to those seen on MDMA and other "ecstasy" tablets. See also Table A10.1.

A10.1.4 Pharmacology

Amphetamine is a central nervous system (CNS) stimulant that causes hypertension and tachycardia with feelings of increased confidence,

Table A10.1 Selected chemical properties of the major drugs of abuse[a].

Name	Molecular formula	Molecular weight of base (Daltons)	Typical salt	Base content of salt
Amphetamine	$C_9H_{13}N$	135.2	Sulfate	73%
Cannabis (THC)	$C_{21}H_{26}O_2$	310.4	n/a	n/a
Cocaine	$C_{17}H_{21}NO_4$	303.4	Hydrochloride	89%
Diamorphine	$C_{21}H_{23}NO_5$	369.4	Hydrochloride	91%
Lysergide (LSD)	$C_{20}H_{25}N_3O$	323.4	Tartrate	78%
MDMA	$C_{11}H_{15}NO_2$	193.2	Hydrochloride	84%
Methylamphetamine	$C_{10}H_{15}N$	149.2	Hydrochloride	80%

[a]The hydrate hydrochloride of diamorphine has a base content of 87%. The phosphate salt of MDMA is also seen

sociability and energy. It suppresses appetite and fatigue and leads to insomnia. Following oral use, the effects usually start within 30 minutes and last for many hours. Later, users may feel irritable, restless, anxious, depressed and lethargic. It increases the activity of the noradrenaline and dopamine neurotransmitter systems. Amphetamine is less potent than methylamphetamine, but in uncontrolled situations, the effects are almost indistinguishable. The S-enantiomer has greater activity than the R-enantiomer. It is rapidly absorbed after oral administration. After a single oral dose of 10 mg, maximum plasma levels are around 0.02 mg/L. The plasma half-life varies from 4 to 12 hours and is dependent on the urinary pH; alkaline urine decreases the rate of elimination. A major metabolite is 1-phenyl-2-propanone with smaller amounts of 4-hydroxyamphetamine. Interpretation of amphetamine in urine is confounded because it is a metabolite of methylamphetamine and certain medicinal products[2]. Acute intoxication causes serious cardiovascular disturbances as well as behavioural problems that include agitation, confusion, paranoia, impulsivity and violence. Chronic use of amphetamine causes neurochemical and neuroanatomical changes, dependence – as shown by increased tolerance, deficits in memory and in decision making and verbal reasoning. Some of the symptoms resemble paranoid schizophrenia. These effects may outlast drug use, although often resolve eventually. Injection of amphetamine carries the same viral infection hazards (*e.g.* HIV and hepatitis) as are found with other injectable drugs such as heroin. Fatalities directly attributed to amphetamine are rare. The estimated minimum lethal dose in nonaddicted adults is 200 mg.

A10.1.5 Synthesis and Precursors

The most common route of synthesis is by the Leuckart method. This uses 1-phenyl-2-propanone (P2P, BMK, phenylacetone) and reagents such as formic acid, ammonium formate or formamide to yield a racemic mixture of the R- and S-enantiomers. A much less common, but stereoselective, method is by reduction of the appropriate diastereoisomers of norephedrine or norpseudoephedrine. Norephedrine and 1-phenyl-2-propanone are listed in Table I of UN1988 (Appendix 5). Caffeine is added to amphetamine at source, but glucose and other sugars are used as subsequent cutting agents.

---

[2] J.T. Cody, *Metabolic precursors to amphetamine and methamphetamine*, For. Sci. Rev., 1993, **5(2)**, 109–127

A10.1.6 Mode of Use

Amphetamine may be ingested, snorted and less commonly injected. Unlike the hydrochloride salt of methylamphetamine, amphetamine sulfate is insufficiently volatile to be smoked. When ingested, a dose may vary from several tens to several hundreds of milligrams of powder depending on the purity.

A10.1.7 Other Names

The term amfetamine (the International Nonproprietary Name: INN) refers to a racemic mixture of the two enantiomers. Amfetamine is also the name required by Directives 65/65 and 92/27/EEC for the labelling of medicinal products within the EU. Dexamfetamine is the INN for the S-α-methylbenzeneethanamine enantiomer, also known as (+)-α-methylphenethylamine. Levamfetamine is the R-α-methylbenzeneethanamine enantiomer, also known as (–)-α-methylphenethylamine. Other commonly used chemical names include: 1-phenyl-2-aminopropane and phenylisopropylamine. Amphetamine is sometimes included with methylamphetamine and other less common substances (*e.g.* benzphetamine) under the generic heading of "amphetamines". Hundreds of other synonyms and proprietary names exist[3]. "Street" terms include speed, base and whizz.

A10.1.8 Analysis

The Marquis field test produces an orange/brown coloration. The Simon test produces a red coloration that will distinguish amphetamine (a primary amine) from secondary amines such as methylamphetamine (blue coloration). The mass spectrum shows little structure with a major ion at $m/z = 44$. Identification by gas-chromatography/mass spectrometry can be improved by N-derivatisation, *e.g.* using carbon disulfide to form the isothiocyanate. Using gas-chromatography, the limit of detection in urine is $< 10\,\mu g/L$.

A10.1.9 Control Status

The R and S-enantiomers (levamfetamine and dexamfetamine, respectively) as well as the racemate (a 50:50 mixture of the R and

---

[3] See for example http://www.chemindustry.com/chemicals/105322.html

Profiles of the Major Drugs of Misuse

S-stereoisomers) are listed in Schedule II of UN1971. In the Misuse of Drugs Act, amphetamine is a Class B controlled drug.

A10.1.10 Medical Use

Amphetamine has occasional therapeutic use in the treatment of narcolepsy and attention deficit hyperactivity disorder (ADHD).

A10.2 CANNABIS

Structure (A10.2) Δ^9-Tetrahydrocannabinol, the major psychoactive principal of cannabis showing the partial ring-numbering system in the more common dibenzopyran system

A10.2.1 Introduction

Cannabis is a natural product, the main psychoactive constituent of which is tetrahydrocannabinol (Δ^9-THC). The cannabis plant (*Cannabis sativa* L.) is broadly distributed and grows in temperate and tropical areas. Together with tobacco, alcohol and caffeine, it is one of the most widely consumed drugs throughout the world, and has been used as a drug and a source of fibre since historical times. Herbal cannabis consists of the dried flowering tops and leaves. Cannabis resin is a compressed solid made from the resinous parts of the plant, and cannabis (hash) oil is a solvent extract of cannabis or cannabis resin. Cannabis is almost always smoked, often mixed with tobacco. Almost all consumption of herbal cannabis and resin is of illicit material. Some therapeutic benefit as an analgesic has been claimed for cannabis, and dronabinol is a licensed medicine in some countries for the treatment of nausea in cancer chemotherapy. Cannabis products and Δ^9-THC are under international control.

A10.2.2 Chemistry

The major active principle in all cannabis products is Δ^9-tetrahydrocannabinol (Δ^9-THC or simply THC; Structure (A10.2)), also known by its International Nonproprietary Name (INN) as dronabinol. The unsaturated bond in the cyclohexene ring is located between C_9 and C_{10} in the more common dibenzopyran ring-numbering system. There are four stereoisomers of THC, but only the (–)-*trans* isomer occurs naturally (CAS-1972-08-03). The fully systematic name for this THC isomer is (–)-(6aR,10aR)-6,6,9-trimethyl-3-pentyl-6a,7,8,10a-tetrahydro-6H-benzo[c]chromen-1-ol. Two related substances, Δ^9-tetrahydrocannabinol-2-oic acid and Δ^9-tetrahydrocannabinol-4-oic acid (both known as THCA) are also present in cannabis, sometimes in large amounts. During smoking, THCA is partly converted to THC. The active isomer Δ^8-THC, where the unsaturated bond in the cyclohexene ring is located between C_8 and C_9, is found in much smaller amounts. Other closely related substances that occur in cannabis include cannabidiol (CBD) and, in aged samples, cannabinol (CBN), both of which have quite different pharmacological effects to THC. Other compounds include the cannabivarins and cannabichromenes; they are all collectively known as cannabinoids. Unlike many psychoactive substances, cannabinoids are not nitrogenous bases.

A10.2.3 Physical Form

Cannabis sativa is dioecious: there are separate male and female plants. The THC is largely concentrated around the flowering parts of the female plant. The leaves and male plants have less THC, while the stalks and seeds contain almost none. Plants have characteristic compound leaves with up to eleven separate serrated lobes. Imported herbal cannabis occurs as compressed blocks of dried brown vegetable matter comprising the flowering tops, leaves, stalks and seeds of *Cannabis sativa*. Cannabis resin is usually produced in 250-g blocks (so-called 9 [ounce] bars), many of which carry a brandmark impression. Cannabis oil is a dark viscous liquid.

A10.2.4 Pharmacology

The pharmacology of cannabis is complicated by the presence of a wide range of cannabinoids. With small doses, cannabis produces euphoria, relief of anxiety, sedation and drowsiness. In some respects, the effects are similar to those caused by alcohol. Anandamide has been identified as the endogenous ligand for the cannabinoid receptor and has pharmacological properties similar to those of THC. When cannabis is smoked, THC can be detected in plasma within seconds of inhalation; it has a half-life of 2 hours. Following smoking of the equivalent of 10–15 mg over a period of 5–7 minutes, peak plasma levels of Δ^9-THC are around 100 µg/L. It is highly lipophilic and widely distributed in the body. Two active metabolites are formed: 11-hydroxy-Δ^9-THC and 8β-hydroxy-Δ^9-THC. The first is further metabolised to Δ^9-THC-11-oic acid. Two inactive substances are also formed: 8α-hydroxy-Δ^9-THC and 8α, 11-dihydroxy-Δ^9-THC and many other minor metabolites, most of which appear in the urine and faeces as glucuronide conjugates. Some metabolites can be detected in the urine for up to two weeks following smoking or ingestion. There is little evidence for damage to organ systems among moderate users, but consumption with tobacco carries all of the risks of that substance. Most interest in the adverse properties of cannabis has centred on its association with schizophrenia, although it is still unclear if there is a causative relation between cannabis use and poor mental health. Fatalities directly attributable to cannabis are rare.

A10.2.5 Origin

Herbal cannabis imported into Europe may originate from West Africa, the Caribbean or South East Asia, but cannabis resin derives largely from either North Africa or Afghanistan. Cannabis oil (hash oil) is often produced locally from cannabis or cannabis resin by means of solvent extraction. Intensive indoor cultivation has become widespread in Europe and elsewhere. This is based on improved seed varieties and procedures such as artificial heating and lighting, hydroponic cultivation in nutrient solutions and propagation of cuttings of female plants. It leads to a high production of flowering material (sometimes known as "sinsemilla" or "skunk") where the THC content may be in excess of 20%.

As with other naturally occurring drugs of misuse (*e.g.* heroin and cocaine) total synthesis is not currently an economic proposition. No precursors to THC are listed in the United Nations 1988 Convention Against Illicit Traffic in Narcotic Drugs and Psychotropic Substances.

A10.2.6 Mode of Use

Cannabis is almost always smoked, often mixed with tobacco either in a cigarette or in a smoking device (bong). Because THC has a low water solubility, ingestion of cannabis leads to poor absorption. The average "reefer" cigarette contains around 200 mg of herbal cannabis or cannabis resin.

A10.2.7 Other Names

In many countries, herbal cannabis and cannabis resin are formally known as marijuana and hashish (or just "hash"), respectively. Cannabis cigarettes may be termed reefers, joints or spliffs. Street terms for cannabis/cannabis resin include bhang, charas, pot, dope, ganja, hemp, weed, blow, grass and many others.

A10.2.8 Analysis

Although the leaves of *Cannabis sativa* are reasonably characteristic, cannabis and cannabis resin can both be positively identified by low-power microscopy, where the appearance of glandular trichomes and cystolithic hairs is diagnostic. The Duquenois test is considered to be specific for cannabinols. It is based on the reaction of cannabis extracts with *p*-dimethylbenzaldehyde (Ehrlich's reagent). This produces a violet blue coloration that is extractable into chloroform. The mass spectrum of THC shows major ions at $m/z = 299, 231, 314, 43, 41, 295, 55$ and 271. Using gas-chromatography, the limit of detection of THC in blood is $0.3\,\mu g/L$.

A10.2.9 Control Status

Cannabis and cannabis resin are both listed in Schedules I and IV of UN1961. In Article 1, Paragraph 1 of that Convention, cannabis is defined as: *"The flowering or fruiting tops of the cannabis plant (excluding the seeds and leaves when not accompanied by the tops) from which the resin has not been extracted, by whatever name they may be designated"*. Cannabis resin is defined as: *"The separated resin, whether crude or purified, obtained from the cannabis plant"*. Along with a number of its isomers and stereochemical variants, Δ^9-THC is listed in Schedule I of UN1971. However, dronabinol is listed in Schedule II. In the Misuse of Drugs Act, cannabis, cannabis resin, cannabinol and derivatives of cannabinol are Class C controlled drugs (Class B from 2009).

A10.2.10 Medical Use

Tinctures of cannabis (ethanolic extracts) were once common, but were removed from Pharmacopoeias many years ago. Herbal cannabis (known as "cannabis flos"), with a nominal THC content of 18% is available as a prescription medicine in the Netherlands. It is indicated for multiple sclerosis, certain types of pain and other neurological conditions. An extract of cannabis (Sativex®) has been licensed in Canada.

A10.3 COCAINE AND CRACK

Structure (A10.3) Cocaine

A10.3.1 Introduction

Cocaine is a natural product extracted from the leaves of *Erythroxylon coca* Lam. This tropical shrub is cultivated widely on the Andean ridge in South America and is the only known natural source of cocaine. Normally produced as the hydrochloride salt, it has limited medical use as a topical anaesthetic. The free base, sometimes known as crack, is a smokable form of cocaine. Coca leaves have been used as a stimulant by some indigenous people of South America since historical times. Purified cocaine has been misused as a central nervous system (CNS) stimulant since the early years of the 20th century. Cocaine is under international control.

A10.3.2 Chemistry

The systematic name is [1R-(exo,exo)]-3-(benzoyloxy)-8-methyl-8-azabicyclo[3.2.1]octane-2-carboxylic acid methyl ester; Structure (A10.3). Cocaine is the methyl ester of benzoylecgonine and is also known as 3β-hydroxy-1αH, 5α-H-tropane-2β-carboxylic acid methyl ester benzoate.

Although four pairs of enantiomers are theoretically possible, only one (commonly termed l-cocaine) occurs naturally. Cocaine is structurally related to atropine (hyoscamine) and hyoscine (scopolamine): substances with quite different pharmacological properties.

A10.3.3 Physical Form

Cocaine base (CAS-50-36-2) and the hydrochloride salt (CAS-53-21-4) are white powders. When in the form of crack, cocaine base usually occurs as small (100–200 mg) lumps known as "rocks".

A10.3.4 Pharmacology

Cocaine shares a similar psychomotor stimulant effect to that seen in amphetamine and related compounds[4]. It increases transmitter concentrations in both the noradrenergic and the dopaminergic synapse and also acts as an anaesthetic agent. Like amphetamine, it produces euphoria, tachycardia, hypertension and appetite suppression. Cocaine has a strong reinforcing action, causing a rapid psychological dependence; an effect even more pronounced in those who smoke cocaine base. Following a 25-mg dose, blood levels peak in the range 400–700 µg/L depending on the route of administration. The main metabolites are benzoylecgonine, ecgonine and ecgonine methyl ester, all of which are inactive. When consumed with alcohol, cocaine also produces the metabolite cocaethylene. Some unchanged cocaine is found in the urine. The plasma half-life of cocaine is 0.7 to 1.5 hours and is dose dependent. The estimated minimal lethal dose is 1.2 g, but susceptible individuals have died from as little as 30 mg applied to mucous membranes, whereas addicts may tolerate up to 5 g daily.

A10.3.5 Origin/Extraction

Dried coca leaves contain up to 1% cocaine[5]. They are processed into cocaine hydrochloride in clandestine laboratories. The leaves are

---

[4] J. Grabowski (ed.), *Cocaine: pharmacology, effects, and treatment of abuse*, National Institute on Drug Abuse (NIDA) Research Monograph 50, 1984

[5] D.A. Cooper, *Clandestine production processes for cocaine and heroin*, in *Clandestinely produced drugs, analogues and precursors: problems and solutions*, ed. M. Klein, F. Sapienza, H. McClain and I. Khan, United States Department of Justice Drug Enforcement Administration, Washington, D.C., 1989

moistened with lime water or other alkali and extracted with kerosene (paraffin). The dissolved cocaine is extracted from the kerosene with sulfuric acid to produce an aqueous solution of cocaine sulfate. This solution is neutralised with lime causing cocaine base (coca paste) to precipitate. Coca paste is redissolved in sulfuric acid and potassium permanganate is added to destroy cinnamoylcocaine and other impurities. The filtered solution is again treated with alkali to precipitate the free base, which is dissolved in acetone or other solvents. Concentrated hydrochloric acid is added to the solution causing cocaine hydrochloride to settle out as a solid residue. Sulfuric and hydrochloric acids, acetone and certain other solvents are listed in Table II, and potassium permanganate is listed in Table I, of UN1988 (Appendix 5).

Although various methods exist for the synthesis of cocaine, they are less economic than extraction of the natural product. Typical precursors include atropine, tropinone and carbomethoxytropinone, none of which is listed in the above-mentioned Convention.

Crack is manufactured from cocaine hydrochloride by one of two main methods: either microwaving a wet mixture with sodium bicarbonate or by adding alkali to a hot saturated solution of cocaine and allowing the denser base to settle and solidify.

A10.3.6 Mode of Use

In illicit use, cocaine is typically snorted (insufflated) where it is absorbed through the nasal mucosa. Ingestion leads to loss of activity through enzymic hydrolysis in the gut. Crack is a smokable form of cocaine. Injection of cocaine is less common. A typical dose of cocaine or crack is 100–200 mg at "street" purity.

A10.3.7 Other Names

Street terms include coke, snow, charlie and a wide variety of others depending on location and setting.

A10.3.8 Analysis

The Marquis field test does not form a coloured product with cocaine. A more satisfactory presumptive test is based on either cobalt thiocyanate (giving a blue coloration) or p-dimethylbenzaldehyde (Ehrlich's reagent) giving a red coloration. Cocaine also produces the characteristic odour of methyl benzoate when heated with a mixture of methanol and sodium hydroxide solution. In the mass spectrum, the major ions are $m/z = 82$, 182, 83, 105, 303, 77, 94 and 96. Using gas-chromatography, the limit of detection in blood is 20 µg/L.

A10.3.9   Control Status

Cocaine is listed in Schedule I of UN1961. The esters and derivatives of ecgonine, which are convertible to ecgonine and cocaine, are also controlled according to that Convention. Coca leaf is separately listed in Schedule I and is defined by Article 1, Paragraph 1 as: *"The leaf of the coca bush, except a leaf from which all ecgonine, cocaine and any other ecgonine alkaloids have been removed"*. In the Misuse of Drugs Act, cocaine is a Class A controlled drug.

A10.3.10   Medical Use

Solutions of cocaine hydrochloride have limited medical use as a topical anaesthetic for surgical procedures involving the eye, ear, nose and throat.

A10.4   DIAMORPHINE (HEROIN)

Structure (A10.4) Diacetylmorphine, the principal psychoactive constituent of heroin

A10.4.1   Introduction

Heroin is a crude preparation of diamorphine. It is a semi-synthetic product obtained by acetylation of morphine, which occurs as a natural product in opium: the dried latex of certain poppy species (*e.g. Papaver somniferum* L). Diamorphine is a narcotic analgesic used in the treatment of severe pain. Illicit heroin may be smoked or solubilised with a weak acid and injected. Whereas opium has been smoked since historical times, diamorphine was first synthesised in the late 19th century. Heroin is under international control.

A10.4.2 Chemistry

Diamorphine (diacetylmorphine; CAS-561-27-3) is produced by the acetylation of morphine. The systematic name is (5α,6α)-7,8-didehydro-4,5-epoxy-17-methylmorphinan-3,6-diol acetate; Structure (A10.4). Although five pairs of enantiomers are theoretically possible in morphine, only one occurs naturally (5R, 6S, 9R, 13S, 14R). Apart from adulterants, crude heroin contains variable amounts of other opium alkaloids (*e.g.* monoacetylmorphine, noscapine, papaverine and acetylcodeine)[6]. The hydrolysis product (6-monoacetylmorphine) may also be present and arises when heroin is stored in damp conditions or in nonacidified aqueous solutions.

A10.4.3 Physical Form

South West Asian heroin is a brown powder usually in the form of the free base, which is insoluble in water, but soluble in organic solvents. The less common South East Asian heroin is usually a white powder containing the hydrate hydrochloride salt of diamorphine (CAS-1502-95-0), which is soluble in water but insoluble in organic solvents.

A10.4.4 Pharmacology

Diamorphine, like morphine and many other opioids, produces analgesia. It behaves as an agonist at a complex group of receptors (the μ, κ and δ subtypes) that are normally acted upon by endogenous peptides known as endorphins. Apart from analgesia, diamorphine produces drowsiness, euphoria and a sense of detachment. Negative effects include respiratory depression, nausea and vomiting, decreased motility in the gastrointestinal tract, suppression of the cough reflex and hypothermia. Tolerance and physical dependence occur on repeated use. Cessation of use in tolerant subjects leads to characteristic withdrawal symptoms. Subjective effects following injection are known as "the rush" and are associated with feelings of warmth and pleasure, followed

---

[6] P.L. Schiff, *Opium and its alkaloids*, Am. J. Pharm. Ed., 2002, **66**, 186–194

by a longer period of sedation. Diamorphine is 2 to 3 times more potent than morphine. The estimated minimum lethal dose is 200 mg, but addicts may be able to tolerate ten times as much. Following injection, diamorphine crosses the blood-brain barrier within 20 seconds, with almost 70% of the dose reaching the brain. It is difficult to detect in blood because of rapid hydrolysis to 6-monoacetylmorphine and slower conversion to morphine, the main active metabolite. The plasma half-life of diamorphine is about 3 minutes. Morphine is excreted in the urine largely as the glucuronide conjugate. Diamorphine is associated with far more accidental overdoses and fatal poisonings than any other scheduled substance. Much morbidity is caused by infectious agents transmitted by unhygienic injection.

A10.4.5 Origin/Extraction

The latex from the seed capsules of the opium poppy (*Papaver somniferum* L.) is allowed to dry[7]. This material (opium) is dispersed in an aqueous solution of calcium hydroxide (slaked lime). The alkalinity is adjusted by adding ammonium chloride causing morphine base to precipitate. The separated morphine is boiled with acetic anhydride. Sodium carbonate is added causing the crude diamorphine base to separate. Depending on the region, this may be used directly, further purified or converted into the hydrochloride salt.

Until the late 1970s, nearly all heroin consumed in Europe came from South East Asia, but now most originates from South West Asia, an area centred on Afghanistan and Pakistan. Heroin is also produced in certain parts of South America and in Mexico (Black Tar Heroin), but those materials are rarely seen in Europe. Acetic anhydride, an essential precursor in the manufacture of heroin is listed in Table I of UN1988 (Appendix 5). As with other naturally occurring drugs of misuse (*e.g.* cocaine and cannabis) total synthesis of the active principals is not currently worthwhile.

A10.4.6 Mode of Use

Heroin from South West Asia may be "smoked" by heating the solid on a metal foil above a small flame and inhaling the vapour. Those intending

---

[7] D.A. Cooper, *Clandestine production processes for cocaine and heroin*, in *Clandestinely produced drugs, analogues and precursors: problems and solutions*, ed. M. Klein, F. Sapienza, H. McClain and I. Khan, United States Department of Justice Drug Enforcement Administration, Washington, D.C., 1989

to inject this form of heroin must first solubilise it with, for example, citric acid or ascorbic acid. Heroin from South East Asia is suitable for direct injection of a solution. A typical dose is 100 mg at street-level purity. Ingestion of diamorphine/heroin is a much less effective route.

A10.4.7 Other Names

A large number of street terms are in use, including horse, smack, shit and brown.

A10.4.8 Analysis

In common with many other opioids, the Marquis field test produces a violet/purple coloration. In the mass spectrum, the major ions are $m/z =$ 327, 43, 369, 268, 310, 42, 215 and 204. Using gas-chromatography, the limits of detection of both diamorphine and 6-monoacetylmorphine are 100 µg/L in body fluids.

A10.4.9 Control Status

Heroin is listed in Schedule I of UN1961. Diamorphine is also included in a generic sense since the 1972 Protocol, which revised the 1961 Convention, extended control to esters or ethers of scheduled substances. Thus, diamorphine is the diacetyl-ester of morphine (Schedule I). In the Misuse of Drugs Act, diamorphine is a Class A controlled drug.

A10.4.10 Medical Use

Diamorphine is a narcotic analgesic with limited use in the treatment of severe pain.

A10.5 LYSERGIDE (LSD)

Structure (A10.5) Lysergide

A10.5.1 Introduction

Lysergide (LSD) is a semi-synthetic hallucinogen, and is amongst the most potent drugs known. Recreational use became popular during the 1960s to 1980s, but is now less common. It is generally believed that most LSD was produced in the US, but the preparation of dosage units by dipping or spotting paper squares was more widespread. These dosage units usually bear coloured designs featuring cartoon characters, geometric and abstract motifs. LSD is related to other substituted tryptamines, and is under international control.

A10.5.2 Chemistry

The International Nonproprietary Name (INN) is (+)-lysergide. The abbreviation LSD derives from **Ly**Sergic acid **D**iethylamide (CAS-50-37-3; Structure (A10.5)). Lysergide belongs to a family of indole alkylamines that includes numerous substituted tryptamines such as psilocin (found in "magic" mushrooms) and N,N-dimethyltryptamine (DMT). The IUPAC name for LSD is 9,10-didehydro-N,N-diethyl-6-methyl-lergoline-8β-carboxamide. The R-stereoisomer has the highest activity.

A10.5.3 Physical Form

LSD is normally produced as the tartrate salt, which is colourless, odourless and water soluble. The common street dose forms are "blotters" or "paper squares" – sheets of absorbent paper printed with distinctive designs and perforated so they may be torn into single 7-mm squares each containing a single dose. Each sheet typically contains 100 or more doses. LSD is less commonly seen as small tablets (microdots) that are 2–3 mm diameter, as thin gelatine squares (window panes) or in capsules. Solutions of LSD in water or alcohol are occasionally encountered. LSD is light sensitive in solution, but more stable in dosage units.

A10.5.4 Pharmacology

LSD was synthesised by Albert Hoffman while working for Sandoz Laboratories in Basel in 1938. Some years later, during a re-evaluation of the compound he accidentally ingested a small amount and described the first "trip". During the 1950s and 1960s, Sandoz evaluated the drug for therapeutic purposes and marketed it under the name Delysid®. It was used for research into the chemical origins of mental illness.

Recreational use started in the 1960s and led to the "psychedelic period".

Physical effects (*e.g.* dilated pupils, mild hypertension and occasionally raised body temperature) appear first. Sensory-perceptual changes are the outstanding features of LSD. Visual disturbances are perceived with eyes closed or open and may consist of geometric shapes and other patterns. Flashes of intense colour are seen and stable objects may appear to move and dissolve. Cross-sensory perception (synaesthesia) such as "coloured hearing" can occur where sounds such as voices or music evoke perception of particular colours or shapes. Time may appear to slow down.

The mode of action of LSD is not well understood. It is thought to interact with the serotonin system by binding to and activating the 5-hydroxytryptamine subtype 2 receptor (5-HT$_2$), which interferes with inhibitory systems resulting in perceptual disturbances It is amongst the most potent drugs known, being active at doses from about 20 µg. Typical doses are now about 20 to 80 µg although in the past, doses as high as 300 µg were common. Like other hallucinogens, dependence does not occur.

When taken orally, the effects become apparent within about 30 minutes and may continue for 8 to 12 hours or more. The duration and intensity of effects are dose dependent. The plasma half-life is about 2.5 hours. Following a dose of 160 µg to 13 subjects, plasma concentrations varied considerably up to 9 µg/L. In humans, LSD is extensively transformed in the liver by hydroxylation and glucuronide conjugation to inactive metabolites. Only about 1% is excreted unchanged in the urine in 24 hours. A major metabolite found in urine is 2-oxylysergide.

Panic reactions (bad trips) may be sufficiently severe to require medical support. Patients usually recover within a few hours but occasionally hallucinations last up to 48 hours and psychotic states for 3–4 days. The effects are greatly affected by the set (an individual's mental state) and the setting (surroundings) in which the drug is taken. Sensory disturbances known as "flashbacks" sometimes occur. Serious side effects often attributed to LSD such as irrational acts leading to suicide or accidental deaths are extremely rare. Deaths attributed to LSD overdose are virtually unknown.

A10.5.5 Synthesis and Precursors

Methods for producing LSD are complex and require an experienced chemist. Several methods are known, but the majority use lysergic acid as the precursor. Lysergic acid itself is also often produced in clandestine laboratories using ergometrine or ergotamine tartrate as the starting material. Ergotamine occurs naturally in the ergot fungus (*Claviceps purpurea*), a common parasite on rye. Depending on the method used, other essential reagents include N,N-carbonyldi-imidazole, diethylamine or hydrazine. Absorbent paper doses (blotters) are prepared by dipping the paper in an aqueous alcoholic solution of the tartrate salt or by dropping the solution on to individual squares.

Ergometrine (also known as ergonovine), ergotamine and lysergic acid are listed in Table I of the United Nations 1988 Convention Against Illicit Traffic in Narcotic Drugs and Psychotropic Substances (see Appendix 5).

A10.5.6 Mode of Use

LSD is taken orally. Paper doses are placed on the tongue, where the drug is rapidly absorbed. Tablets or capsules are swallowed. LSD is not absorbed through dry skin.

A10.5.7 Other Names

Synonyms include N,N-diethyl-lysergamide, lysergic acid diethylamide, LSD, and LSD-25. There are many street names including acid, blotter, dots, tabs, tickets, trips and many others related to the particular designs on the paper dosage forms.

A10.5.8 Analysis

LSD may be detected in paper doses after extracting the drug into methanol. The extract is spotted onto filter paper, dried and examined under ultraviolet light (360 nm); LSD gives a strong blue fluorescence. Ehrlich's reagent (*p*-dimethyl-aminobenzaldehyde) gives a blue/purple colour and may be applied after thin layer chromatography. HPLC with fluorescence detection or gas-chromatography/mass spectrometry are used for confirmation or quantification. The major ions in the mass spectrum are $m/z = 323, 221, 181, 222, 207, 72, 223$ and 324. Commercial immunoassays are available for the detection of LSD in urine at concentrations at or above $0.5\,\mu g/L$.

Many ergot alkaloids can interfere with LSD analysis, *e.g.* ergometrine, methylergometrine, dihydroergotamine, ergocornine, ergocristine, methysergide, and ergotamine. LSD degrades readily, particularly in biological specimens, unless protected from light and elevated temperatures; it may also bind to glass containers in acidic solutions. The only analogues of LSD to have received widespread interest are the *N*-methylpropylamide (LAMPA) and the *N*-butylamide of lysergic acid. It seems that neither has appeared in seized material, but both are regarded more as theoretical interferences. It is assumed that any analytical technique should be capable of separating them[8] from LSD.

A10.5.9 Control Status

LSD is listed in Schedule I of the United Nations 1971 Convention on Psychotropic Substances. In the Misuse of Drugs Act, lysergide is a Class A controlled drug.

A10.5.10 Medical Use

Although once used in psychotherapy, LSD has no current medical use.

A10.6 MDMA

Structure (A10.6) 3,4-Methylenedioxymethylamphetamin

A10.6.1 Introduction

MDMA is a synthetic substance commonly known as ecstasy, although the latter term has now been generalised to cover a wide range of other substances. Originally developed around 1912 by the Merck chemical company, it was never marketed. Although proposed as an aid to psychiatric counselling, therapeutic use is extremely limited. Illicit MDMA is normally seen as tablets, many of which are manufactured in Europe.

---

[8] C.C. Clark, *The differentiation of lysergic acid diethylamide (LSD) from N-methyl-N-propyl and N-butyl amides of lysergic acid.* J. For. Sci., 1989, **34(3)**; published online at: http://journal-sip.astm.org/JOURNALS/FORENSIC/PAGES/662.htm

It acts as a central nervous system (CNS) stimulant and has a weak hallucinogenic property more accurately described as increased sensory and social awareness (entactogenicity and empathogenicity). MDMA is under international control.

A10.6.2 Chemistry

MDMA is an acronym for **M**ethylene**D**ioxy**M**ethyl**A**mphetamine; Structure (A10.6). The formal name is *N*-methyl-1-(3,4-methylenedioxyphenyl)propan-2-amine, but MDMA (CAS-42542-10-09) is commonly known as 3,4-methylenedioxymethamphetamine or methylenedioxymethylamfetamine. Other chemical names include N,α-dimethyl-3,4-methylenedioxyphenethylamine or, less usually, *N*-methyl-1-(1,3-benzodioxol-5-yl)-2-propanamine. A number of homologous compounds with broadly similar effects, *e.g.* MDA (**M**ethylene**D**ioxy-**A**mphetamine), MDEA (**M**ethylene**D**ioxy**E**thyl**A**mphetamine) and MBDB [*N*-**M**ethyl-1-(1,3-**B**enzo**D**ioxol-5-yl)-2-**B**utanamine] have appeared, but have proved less popular. These substances and certain other ring-substituted phenethylamines are collectively known as the ecstasy drugs. As with other phenethylamines, and like its close relative methylamphetamine, MDMA also exists in two enantiomeric forms (*R* and *S*).

A10.6.3 Physical Form

The most common salt is the hydrochloride (CAS-64057-70-1): a white or off-white powder or crystals soluble in water. The phosphate salt is also encountered. Illicit products are seen principally as white tablets with a characteristic impression (logo), less commonly as powders or capsules. MDMA base is a colourless oil insoluble in water.

A10.6.4 Pharmacology

Whereas phenethylamines without ring substitution usually behave as stimulants, ring-substitution (as in MDMA) leads to a modification in the pharmacological properties. Ingestion of MDMA causes euphoria, increased sensory awareness and mild central stimulation. It is less hallucinogenic than its lower homologue, methylenedioxyamphetamine (MDA). The terms empathogenic and entactogenic have been coined to describe the socialising effects of MDMA. Following ingestion, most of the dose of MDMA is excreted in the urine unchanged. Major metabolites are 3,4-methylenedioxyamphetamine (MDA) and *O*-demethylated compounds. Using a dose of 75 mg, the maximum plasma concentration of around 0.13 mg/L is reached within 2 hours. The plasma half-life is 6–7 hours. In animals, MDMA shows neurotoxicity as evidenced by anatomical changes in axon structure and a persisting reduction in brain serotonin levels. The significance of these findings to human users is still unclear, although cognitive impairment is associated with MDMA use. Some of the pharmacodynamic and toxic effects of MDMA vary depending on which enantiomer is used. However, almost all illicit MDMA exists as a racemic mixture. Fatalities following a dose of 300 mg have been noted, but toxicity depends on many factors including individual susceptibility and the circumstances in which MDMA is used.

A10.6.5 Synthesis and Precursors

There are four principal precursors, which can be used in the manufacture of MDMA and related drugs: safrole, isosafrole, piperonal and 3,4-methylenedioxyphenyl-2-propanone (PMK). Safrole is the key starting material insofar as the other three can be readily synthesised from it. In the original Merck patent of 1914, safrole was reacted with hydrobromic acid to form bromosafrole, which was converted to MDMA using methylamine. Many illicit syntheses start with PMK and use either the Leuckart route or various reductive aminations including the aluminium foil method. All of these methods produce racemic MDMA. The four precursors noted above are listed in Table I of UN1988 (Appendix 5).

A10.6.6 Mode of Use

MDMA is always ingested. Users may consume several tablets in a session.

A10.6.7 Other Names

As some of the above names suggest, MDMA is a derivative of amphetamine and a member of the phenethylamine family. Street terms for MDMA include Adam and XTC, but often reflect the imprinted logo, *e.g.* Mitsubishis, Love Doves and many others.

A10.6.8 Analysis

In common with many of its homologues, MDMA reacts with the Marquis field test to produce a dark blue/black coloration. The mass spectrum shows limited structure with a major ion at m/z = 58 and other ions at m/z = 135 and 77. Using gas-chromatography, the limits of detection in plasma and urine are 1.6 µg/L and 47 µg/L, respectively.

A10.6.9 Control Status

MDMA, shown as (±)-N,α-dimethyl-3,4-(methylenedioxy)phenethylamine, is listed in Schedule I of UN1971. In the Misuse of Drugs Act, MDMA is covered by the generic definition of a substituted phenethylamine: a Class A controlled drug.

A10.6.10 Medical Use

MDMA once found limited use in psychiatric counselling, but its therapeutic use is now rare.

A10.7 METHYLAMPHETAMINE

Structure (A10.7) Methylamphetamine

A10.7.1 Introduction

Methylamphetamine (methamphetamine) is the most widely abused synthetic psychotropic drug, particularly in North America and countries of the Far East. In Europe, it is most commonly reported in the Czech Republic and Slovakia, but is much less widespread in countries where illicit amphetamine is the established stimulant[9]. Normally seen as a white powder, it is a synthetic substance that acts as a stimulant of the

---

[9] P. Griffiths, V. Mravcik, D. Lopez and D. Klempova, *Quite a lot of smoke but very limited fire – the use of methamphetamine in Europe*, Drug and Alcohol Rev., 2008, **27**, 236–242

central nervous system (CNS). First manufactured in Japan in 1919, methylamphetamine has some limited therapeutic use, but most is manufactured in clandestine laboratories. It is under international control and is closely related to amphetamine[10].

A10.7.2 Chemistry

Methylamphetamine (CAS–537-46-2; Structure (A10.7)) is a member of the phenethylamine family, which includes a range of substances that may be stimulants, entactogens or hallucinogens. Thus, methylamphetamine is N, α-dimethylphenethylamine. The fully systematic name is N, α-dimethylbenzeneethanamine. The asymmetric α-carbon atom gives rise to two enantiomers. These two forms were previously called the (−) or l-stereoisomer and the (+) or d-stereoisomer, but in modern usage are defined as the R and S stereoisomers.

A10.7.3 Physical Form

Methylamphetamine base is a colourless volatile oil insoluble in water. The most common salt is the hydrochloride (CAS-51-57-0): a white or off-white powder or crystals soluble in water. Illicit products mostly consist of powders, but the pure crystalline hydrochloride, known as "Ice", is rarely seen outside the Far East. Tablets containing methylamphetamine may carry logos similar to those seen on MDMA and other ecstasy tablets.

A10.7.4 Pharmacology

Methylamphetamine is a central nervous system stimulant that causes hypertension and tachycardia with feelings of increased confidence, sociability and energy[11]. It suppresses appetite and fatigue and leads to insomnia. Following oral use, the effects usually start within 30 minutes and last for many hours. Later, users may feel irritable, restless, anxious, depressed and lethargic. It increases the activity of the noradrenaline

---

[10] M.R. Hammer, *A Key to methamphetamine-related literature*, New York State Department of Health, 2006: http://www.nyhealth.gov/diseases/aids/harm_reduction/crystalmeth/docs/meth_literature_index.pdf

[11] C.E. Cook, A.R. Jeffcoat, J.M. Hill, D.E. Pugh, P.K. Patetta, B.M. Sadler, W.R. White, and M. Perez-Reyes, *Pharmacokinetics of methamphetamine self-administered to human subjects by smoking S-(+)-methamphetamine hydrochloride*, Drug Metab. Dispos., 1993, **21**, 717–723

and dopamine neurotransmitter systems. Methylamphetamine is more potent than amphetamine, but in uncontrolled situations, the effects are almost indistinguishable. The *S*-stereoisomer has greater activity than the *R*-stereoisomer. The therapeutic dose of the *S*-stereoisomer is up to 25 mg orally. It is rapidly absorbed after oral administration, and maximum plasma levels are in the range 0.001 to 0.005 mg/L. The plasma half-life is about 9 hours. The major metabolites include 4-hydroxymethylamphetamine and amphetamine. Fatalities directly attributed to methylamphetamine are rare. In most fatal poisonings the blood concentration is above 0.5 mg/L. As with amphetamine, the interpretation of methylamphetamine in urine is confounded because it is a metabolite of certain medicinal products (*e.g.* selegiline)[12]. Acute intoxication causes serious cardiovascular disturbances as well as behavioural problems that include agitation, confusion, paranoia, impulsivity and violence. Chronic use of methylamphetamine causes neurochemical and neuroanatomical changes, dependence (as shown by increased tolerance), deficits in memory and in decision making and verbal reasoning. Some of the symptoms resemble paranoid schizophrenia. These effects may outlast drug use, although they often resolve eventually. Injection of methylamphetamine carries the same viral infection hazards (*e.g.* HIV and hepatitis) as are found with other injectable drugs such as heroin. When methylamphetamine is smoked it reaches the brain much more quickly. Drugs that are smokable (*e.g.* methylamphetamine, crack cocaine) are much more addictive and more likely to cause problems than when used orally.

A10.7.5 Synthesis and Precursors

The *S*-enantiomer is most commonly produced by reduction of l-ephedrine, *i.e.* (1*R*, 2*S*)-2-methylamino-1-phenylpropan-1-ol, or by reduction of d-pseudoephedrine, *i.e.* (1*S*, 2*S*)-2-methylamino-1-phenylpropan-1-ol. Both ephedrine and pseudoephedrine are commercially available and are used in certain medicinal products. Ephedrine may also be extracted from the plant *Ephedra vulgaris* L. (used in Chinese medicine as Ma Huang). Both the Leuckart route and reductive amination (*e.g.* the aluminium foil method) of 1-phenyl-2-propanone (P2P, BMK, phenylacetone) yield a racemic mixture of the *R*- and *S*-enantiomers. The synthetic route used

---

[12] J.T. Cody, *Metabolic precursors to amphetamine and methamphetamine*, For. Sci. Rev., 1993, **5(2)**, 109–127

may be identified by impurity profiling[13]. Ephedrine, pseudoephedrine and 1-phenyl-2-propanone are listed in Table I of UN1988 (Appendix 5).

A10.7.6 Mode of Use

Methylamphetamine may be ingested, snorted, injected or smoked. Unlike the sulfate salt of amphetamine, methylamphetamine hydrochloride is sufficiently volatile to be smoked. When ingested, a dose may vary from several tens to several hundreds of milligrams depending on the purity and the isomeric composition.

A10.7.7 Other Names

The term metamfetamine (the International Nonproprietary Name: INN) strictly relates to the specific enantiomer S-N, α-dimethylbenzeneethanamine. Metamfetamine is also the name required by Directives 65/65 and 92/27/EEC for the labelling of medicinal products within the EU. In the UK and some other countries, the name used in drugs legislation is methylamphetamine. Other commonly used chemical names include: N-methylamphetamine, 1-phenyl-2-methylaminopropane, phenylisopropylmethylamine, and desoxyephedrine. Methylamphetamine, as the N-methyl derivative of amphetamine, is sometimes included with amphetamine and other less common substances (*e.g.* benzphetamine) under the generic heading of "amphetamines". Hundreds of other synonyms and proprietary names exist[14]. "Street" terms include speed, crank, meth, crystal meth, pervitin (particularly Eastern Europe; a name derived from an earlier medicinal product), yaba and shabu (certain countries in the Far East).

A10.7.8 Analysis

The Marquis field test produces an orange/brown coloration. The Simon test (for secondary amines) produces a blue coloration that will distinguish methylamphetamine from primary amines such as amphetamine (red coloration). In the mass spectrum, the major ions are

---

[13] B. Remberg and A.H. Stead, *Drug Characterization/impurity Profiling, with special Focus on Methylamphetamine: Recent Work of the United Nations International Drug Control Programme,* Bulletin on Narcotics, 1999, **11(1–2)**, 97–117
[14] See for example: http://www.chemindustry.com/chemicals/55866.html

$m/z = 58$, 91, 59, 134, 65, 56, 42, 57. Identification by gas-chromatography/mass spectrometry can be improved by N-derivatisation. Using gas-chromatography, the limit of detection in urine is $< 10\,\mu g/L$.

A10.7.9 Control Status

The S-enantiomer is listed as metamfetamine in Schedule II of UN1971. The racemate (a 50:50 mixture of the R and S-stereoisomers) is also listed in the same Schedule (as metamfetamine racemate), but the R-enantiomer is not separately identified in the Convention. In the Misuse of Drugs Act, methylamphetamine is a Class A controlled drug.

A10.7.10 Medical Use

Methylamphetamine has occasional therapeutic use in the treatment of narcolepsy and attention deficit hyperactivity disorder (ADHD).

APPENDIX 11
Field Tests and the "Guilty Plea Policy"

The number of people arrested in the UK for drug offences has increased rapidly during the past twenty years. Several measures were introduced to ease the burden on and cost to the criminal justice system. In addition to the cautioning of minor offenders, a policy was introduced whereby some of those found to be in possession of suspected controlled substances could be prosecuted without the need for a laboratory identification of those substances. This became known as the "guilty plea policy". Following an arrest, and if the quantities of drugs found were consistent with personal use, the arrestee admitted possession and the substance was claimed to be either cannabis or cannabis resin, then a suitably trained police officer was allowed to identify the seized material by visual inspection. If the substance was claimed to be amphetamine, cocaine, morphine or heroin, then a police officer or other trained person was allowed to use one of a range of Home Office approved drug testing kits. If the claimed identity of the drug in question was confirmed by the test result then that evidence would be acceptable in a Magistrate's Court. In all other cases, a normal laboratory examination of seized materials is required. The policy[1,2] is set out in Home Office Circular 40/1998, modified by Circular 10/2005.

The test for amphetamine, morphine or heroin was based on the reaction of the substance with Marquis reagent (10% formaldehyde in

---

[1] http://www.knowledgenetwork.gov.uk/HO/circular.nsf/79755433dd36a66980256d4f004d1514/bddce48a0361e03580256fb30033bb42?OpenDocument
[2] http://www.nationalarchives.gov.uk/ERORecords/HO/415/1/circulars/1998/hoc9840.htm

Forensic Chemistry of Substance Misuse: A Guide to Drug Control
By L.A. King
© L.A. King 2009
Published by the Royal Society of Chemistry, www.rsc.org

concentrated sulfuric acid). If this produced an orange/brown colour then that was considered to confirm an admission of amphetamine. If the test produced a purple colour then that confirmed morphine or heroin. The test for cocaine involves an immunoassay carried out using a disposable device[3].

Despite the fact that the Marquis test is fairly crude – many other substances give an orange/brown reaction and most opiates give a purple reaction – the guilty plea policy has worked well. However, if the use of methylamphetamine were to become more widespread then this could expose a policy weakness. Thus, methylamphetamine gives the same colour reaction as amphetamine. But as methylamphetamine is now a Class A controlled drug, it might be expected that users would always admit that they had possession of amphetamine (Class B).

Discussions are currently underway between the Association of Chief Police Officers (ACPO) and the National Policing Improvement Agency (NPIA) to consider extending field testing to a wider range of circumstances, *i.e.* guilty and not-guilty pleas, and a wider range of drugs[4]. The motive for this is to reduce the cost and time spent on laboratory testing.

---

[3] Cocaine test kits are manufactured by Cozart Biosciences, Abingdon, Oxfordshire, UK and by Dtec International Ltd., Lytham St. Annes, Lancashire, UK

[4] A trial of drug testing kits in all possession cases (*i.e.* guilty and not-guilty pleas) has been piloted in three UK police forces. It is expected to be adopted throughout England and Wales.

APPENDIX 12
Purities and Drug Content of Illicit Substances

Table A12.1 shows typical (*i.e.* modal values) of drug purity. Apart from amphetamine, modal values are close to mean values (averages) and always refer to the base content of drug or the THC content (potency) for cannabis products. There has been a slow decline in the purities of cocaine and crack and the drug content of MDMA tablets since 1999, but the purity of heroin and amphetamine has shown no clear trend in this time. LSD is now rarely seen and even less commonly quantified; the drug content of paper squares noted here was measured over ten years ago[1].

In some EU countries, there is licensed cultivation of cannabis for the production of hemp fibre, but the THC content of these plants is less than 0.3%. The average "reefer" cigarette contains around 200 mg of herbal cannabis or cannabis resin[2]. Small amounts of cannabis resin are illicitly produced in the EU (*e.g.* nederhasj). Like cannabis oil (hash oil), this product may have a THC content in excess of 30%. As crude and inhomogeneous preparations, the concept of adulteration in cannabis and cannabis resin is less meaningful than for powdered drugs, particularly if the adulterant is other vegetable matter. However, in the past

---

[1] L.A. King, *Drug content of powders and other illicit preparations in the UK*, For. Sci. Int., 1997, **85(2)**, 135–147
[2] L.A. King, C. Carpentier and P. Griffiths, *An Overview of Cannabis Potency in Europe*, Insights, No. 6, EMCDDA, June 2004

Table A12.1 Typical purities/potencies/drug content of illicit substances and common adulterants/cutting agents.

Drug	Street purity/potency	Importation purity/potency	Common adulterants
Amphetamine	8%	30–40%	Caffeine, lactose, paracetamol
Cannabis – herbal (sinsemilla/skunk)	13%	13%	n/a
Cannabis – herbal (traditional)	5%	5%	n/a
Cannabis – resin	5%	5%	n/a
Cocaine	40%	70%	Benzocaine, phenacetin, lignocaine, caffeine, diltiazem, paracetamol, procaine, dimethylterephthalate
Crack cocaine	50%	80%	
Heroin	50%	50%	Caffeine, paracetamol, diazepam, phenacetin, phenobarbitone
Lysergide (LSD)	45 µg/unit	45 µg/unit	n/a
MDMA	70 mg/tablet	70 mg/tablet	Lactose, stearates, talc
Methylamphetamine	~10%	unknown	Caffeine

few years deliberate contamination of herbal cannabis with minute glass beads[3] or lead particles[4] has been recorded. As with powdered drugs, the objective of adulteration is to dilute the active material, although in the case of glass beads this may have also been to improve the appearance of otherwise poor-quality herbal cannabis.

---

[3] *Alert - contamination of herbal or "skunk-type" cannabis with glass beads,* Department of Health, 16 January 2007, London: http://www.info.doh.gov.uk/doh/embroadcast.nsf/vwDiscussionAll/297D9740D0412C9D802572650050A4A0?OpenDocument

[4] F. Busse, L. Omidi, A. Leichtle, M. Windgassen, E. Kluge and M. Stumvoll, *Lead poisoning due to adulterated marijuana,* New England J. Med., 2008, **358(15)**, 1642–1642

APPENDIX 13
Prices and Wrap Sizes of Illicit Drugs

Table A13.1 Typical UK street prices[a] in 2006 and wrap sizes of illicit drugs.

Drug	Wrap size	Price per gram
Amphetamine	0.5–1.0g	£9
Cannabis – herbal	1–4g	£2–£3
Cannabis – resin	1–4g	£2–£3
Cocaine	0.2–0.4g	£49
Crack cocaine	0.1–0.2g	£90
Ecstasy (MDMA, *etc.*)	n/a	£3 (per tablet)
Heroin	0.1–0.3g	£52
Lysergide (LSD)	n/a	£3 (paper square)

[a]*United Kingdom Drug Situation, 2007 Edition, UK Focal Point on Drugs, Annual Report to the European Monitoring Centre for Drugs and Drug Addiction* – see Bibliography

APPENDIX 14
Useful Websites

Table A14.1 Websites for the major agencies concerned with drug control, public policy, drug information or related areas.

Address	Host Organisation
http://eldd.emcdda.europa.eu/	European legal database on drugs
http://www.emea.europa.eu/	European Medicines Agency
http://www.drugabuse.gov	National Institute on Drug Abuse (NIDA)
http://www.drugscope.org.uk	Drugscope
http://www.emcdda.europa.eu/	European Monitoring Centre for Drugs and Drug Addiction (EMCDDA)
http://www.opsi.gov.uk/legislation/about_legislation.htm	All UK legislation since 1988
http://www.homeoffice.gov.uk/	Home Office
http://www.rpsgb.org.uk/	The Royal Pharmaceutical Society of Great Britain
http://www.incb.org/	UN International Narcotics Control Board
http://www.unodc.org/	UN Office of Drugs and Crime
http://www.usdoj.gov/dea	Drug Enforcement Administration (US)
http://www.whitehousedrugpolicy.gov	Office of National Drug Control Policy (US)
http://www.erowid.org/	Drug use information
http://www.tdpf.org.uk/MediaNews_PressReleases_01_03_06.htm	Transform Drug Policy Foundation
http://www.idpc.info/	International Drug Policy Consortium
http://www.ukdpc.org.uk/index.shtml	UK Drug Policy Commission
http://www.dailydose.net/	Drug news digests

Forensic Chemistry of Substance Misuse: A Guide to Drug Control
By L.A. King
© L.A. King 2009
Published by the Royal Society of Chemistry, www.rsc.org

APPENDIX 15

The Misuse of Drugs Act – Schedule 2 (Parts I to III)

The following Tables and subordinate text are set out with paragraph headings as they appear in Schedule 2 to the Act. Tables 15.1, 15.2 and 15.3, respectively, list Class A controlled drugs in Part I of Schedule 2, Class B controlled drugs in Part II and Class C controlled drugs in Part III. Substances or products that are listed in UN1961 or UN1971 are indicated along with the corresponding Schedule of the Regulations (see also Appendices 3 to 5). Where a substance or generic definition was introduced into the Act by a subsequent Modification Order (see Appendix 1), the corresponding Statutory Instrument Number (S.I.) and date are shown.

Table A15.1 Class A controlled drugs (Part I of Schedule 2).

1. The following substances and products, namely:

(a)

Substance or product	UN Convention and Schedule	Modification Order	Schedule in Regulations
Acetorphine	UN1961 (I)		2
Alfentanil	UN1961 (I)	(S.I. 859)1984	2
Allylprodine	UN1961 (I)		2
Alphacetylmethadol	UN1961 (I)		2
Alphameprodine	UN1961 (I)		2
Alphamethadol	UN1961 (I)		2
Alphaprodine	UN1961 (I)		2

Forensic Chemistry of Substance Misuse: A Guide to Drug Control
By L.A. King
© L.A. King 2009
Published by the Royal Society of Chemistry, www.rsc.org

Table A15.1 (*Continued*).

Substance or product	UN Convention and Schedule	Modification Order	Schedule in Regulations
Anileridine	UN1961 (I)		2
Benzethidine	UN1961 (I)		2
Benzylmorphine (3-benzylmorphine)	UN1961 (I)		2
Betacetylmethadol	UN1961 (I)		2
Betameprodine	UN1961 (I)		2
Betamethadol	UN1961 (I)		2
Betaprodine	UN1961 (I)		2
Bezitramide	UN1961 (I)		2
Bufotenine	Not listed		1
Carfentanil	Not listed	(S.I. 2230)1986	2
Clonitazene	UN1961 (I)		2
Coca leaf	UN1961 (I)		1
Cocaine	UN1961 (I)		2
Desomorphine	UN1961 (I)		2
Dextromoramide	UN1961 (I)		2
Diamorphine	UN1961 (I)		2
Diampromide	UN1961 (I)		2
Diethylthiambutene	UN1961 (I)		2
Difenoxin [1-(3-cyano-3,3-diphenylpropyl)-4-phenylpiperidine-4-carboxylic acid]	UN1961 (I)	(S.I. 421)1975	2
Dihydrocodeinone O-Carboxymethyloxime	UN1961 (I)		2
Dihydroetorphine	UN1961 (I)	(S.I. 1243)2003	2
Dihydromorphine	UN1961 (I)		2
Dimenoxadole	UN1961 (I)		2
Dimepheptanol	UN1961 (I)		2
Dimethylthiambutene	UN1961 (I)		2
Dioxaphetyl butyrate	UN1961 (I)		2
Diphenoxylate	UN1961 (I)		2
Dipipanone	UN1961 (I)		2
Drotebanol (3,4-dimethoxy-17-methylmorphinan-6β,14-diol)	UN1961 (I)	(S.I. 771)1973	2
Ecgonine, and any derivative of ecgonine which is convertible to ecgonine or to cocaine	UN1961 (I)		2
Ethylmethylthiambutene	UN1961 (I)		2
Eticyclidine	UN1971 (I)	(S.I. 859)1984	1
Etonitazene	UN1961 (I)		2
Etorphine	UN1961 (I)		2
Etoxeridine	UN1961 (I)		2
Etryptamine	UN1971 (I)	(S.I. 750)1998	1
Fentanyl	UN1961 (I)		2
Furethidine	UN1961 (I)		2
Hydrocodone	UN1961 (I)		2
Hydromorphinol	UN1961 (I)		2
Hydromorphone	UN1961 (I)		2
Hydroxypethidine	UN1961 (I)		2

Table A15.1 (*Continued*).

Substance or product	UN Convention and Schedule	Modification Order	Schedule in Regulations
Isomethadone	UN1961 (I)		2
Ketobemidone	UN1961 (I)		2
Levomethorphan	UN1961 (I)		2
Levomoramide	UN1961 (I)		2
Levophenacylmorphan	UN1961 (I)		2
Levorphanol	UN1961 (I)		2
Lofentanil	Not listed	(S.I. 2230)1986	2
Lysergamide	Not listed		1
Lysergide and other *N*-alkyl derivatives of lysergamide	UN1971 (I)		1
Mescaline	UN1971 (I)		1
Metazocine	UN1961 (I)		2
Methadone	UN1961 (I)		2
Methadyl acetate	UN1961 (I)		2
Methylamphetamine	UN1971 (II)	(S.I. 3331)2006	2
Methyldesorphine	UN1961 (I)		2
Methyldihydromorphine (6-methyldihydromorphine)	UN1961 (I)		2
Metopon	UN1961 (I)		2
Morpheridine	UN1961 (I)		2
Morphine	UN1961 (I)		2
Morphine methobromide, morphine *N*-oxide and other pentavalent nitrogen morphine derivatives	UN1961 (I)		2
Myrophine	UN1961 (I)		2
Nicomorphine (3,6-dinicotinoylmorphine)	UN1961 (I)		2
Noracymethadol	UN1961 (I)		2
Norlevorphanol	UN1961 (I)		2
Normethadone	UN1961 (I)		2
Normorphine	UN1961 (I)		2
Norpipanone	UN1961 (I)		2
Opium, whether raw, prepared or medicinal	UN1961 (I)		1 (raw opium) 2 (medicinal opium)
Oxycodone	UN1961 (I)		2
Oxymorphone	UN1961 (I)		2
Pethidine	UN1961 (I)		2
Phenadoxone	UN1961 (I)		2
Phenampromide	UN1961 (I)		2
Phenazocine	UN1961 (I)		2
Phencyclidine	UN1971 (II)	(S.I. 299)1979	2
Phenomorphan	UN1961 (I)		2
Phenoperidine	UN1961 (I)		2
Piminodine	UN1961 (I)		2
Piritramide	UN1961 (I)		2
Poppy-straw and concentrate of poppy-straw	UN1961 (I) (as concentrate)		1 (as concentrate)

Table A15.1 (*Continued*).

Substance or product	UN Convention and Schedule	Modification Order	Schedule in Regulations
Proheptazine	UN1961 (I)		2
Properidine (1-methyl-4-phenylpiperidine-4-carboxylic acid isopropyl ester)	UN1961 (I)		2
Psilocin	UN1971 (I)		1
Racemethorphan	UN1961 (I)		2
Remifentanil	UN1961 (I)	(S.I.1243)2003	2
Racemoramide	UN1961 (I)		2
Racemorphan	UN1961 (I)		2
Rolicyclidine	UN1971 (I)	(S.I. 859)1984	1
Sufentanil	UN1961 (I)	(S.I. 765)1983	2
Tenocyclidine	UN1971 (I)	(S.I. 859)1984	1
Thebacon	UN1961 (I)		2
Thebaine	UN1961 (I)		2
Tilidate	UN1961 (I)	(S.I. 765)1983	2
Trimeperidine	UN1961 (I)		2
4-Bromo-2,5-dimethoxy-α-methylphenethylamine	UN1971 (I)	(S.I. 421)1975	1
4-Cyano-2-dimethylamino-4,4-diphenylbutane	UN1961 (I)		2
4-Cyano-1-methyl-4-phenylpiperidine	UN1961 (I)		2
N,N-Diethyltryptamine	UN1971 (I)		1
N,N-Dimethyltryptamine	UN1971 (I)		1
2,5-Dimethoxy-α,4-dimethylphenethylamine	UN1971 (I)		1
N-Hydroxy-tenamphetamine	UN1971 (I)	(S.I. 2589)1990	1
2-Methyl-3-morpholino-1,1-diphenylpropanecarboxylic acid	UN1961 (I)		2
4-Methyl-aminorex	UN1971 (I)	(S.I. 2589)1990	1
1-Methyl-4-phenylpiperidine-4-carboxylic acid	UN1961 (I)		2
4-Phenylpiperidine-4-carboxylic acid ethyl ester	UN1961 (I)		2

(b) any compound (not being a compound for the time being specified in sub-paragraph (a) above) structurally derived from tryptamine or from a ring-hydroxy tryptamine by substitution at the nitrogen atom of the side-chain with one or more alkyl substituents but no other substituent; (S.I. 1243)1977

(ba) the following phenethylamine derivatives, namely
[This note is not part of the published Schedule, but apart from 4-MTA, which is now included in Schedule 1 of UN1971, all others are unlisted in UN Conventions; all are (S.I. 3932)2001 and Schedule 1 in the Regulations]

Allyl(α-methyl-3,4-methylenedioxyphenethyl)amine
2-Amino-1-(2,5-dimethoxy-4-methylphenyl)ethanol
2-Amino-1-(3,4-dimethoxyphenyl)ethanol

Table A15.1 (*Continued*).

Benzyl(α-methyl-3,4-methylenedioxyphenethyl)amine
4-Bromo-β,2,5-trimethoxyphenethylamine
N-(4-sec-Butylthio-2,5-dimethoxyphenethyl)hydroxylamine
Cyclopropylmethyl(α-methyl-3,4-methylenedioxyphenethyl)amine
2-(4,7-Dimethoxy-2,3-dihydro-1*H*-indan-5-yl)ethylamine
2-(4,7-Dimethoxy-2,3-dihydro-1*H*-indan-5-yl)-1-methylethylamine
2-(2,5-Dimethoxy-4-methylphenyl)cyclopropylamine
2-(1,4-Dimethoxy-2-naphthyl)ethylamine
2-(1,4-Dimethoxy-2-naphthyl)-1-methylethylamine
N-(2,5-Dimethoxy-4-propylthiophenethyl)hydroxylamine
2-(1,4-Dimethoxy-5,6,7,8-tetrahydro-2-naphthyl)ethylamine
2-(1,4-Dimethoxy-5,6,7,8-tetrahydro-2-naphthyl)-1-methylethylamine
α,α-Dimethyl-3,4-methylenedioxyphenethylamine
α,α-Dimethyl-3,4-methylenedioxyphenethyl(methyl)amine
Dimethyl(α-methyl-3,4-methylenedioxyphenethyl)amine
N-(4-Ethylthio-2,5-dimethoxyphenethyl)hydroxylamine
4-Iodo-2,5-dimethoxy-α-methylphenethyl(dimethyl)amine
2-(1,4-Methano-5,8-dimethoxy-1,2,3,4-tetrahydro-6-naphthyl)ethylamine
2-(1,4-Methano-5,8-dimethoxy-1,2,3,4-tetrahydro-6-naphthyl)-1-methylethylamine
2-(5-Methoxy-2,2-dimethyl-2,3-dihydrobenzo[*b*]furan-6-yl)-1-methylethylamine
2-Methoxyethyl(α-methyl-3,4-methylenedioxyphenethyl)amine
2-(5-Methoxy-2-methyl-2,3-dihydrobenzo[*b*]furan-6-yl)-1-methylethylamine
β-Methoxy-3,4-methylenedioxyphenethylamine
1-(3,4-Methylenedioxybenzyl)butyl(ethyl)amine
1-(3,4-Methylenedioxybenzyl)butyl(methyl)amine
2-(α-Methyl-3,4-methylenedioxyphenethylamino)ethanol
α-Methyl-3,4-methylenedioxyphenethyl(prop-2-ynyl)amine
N-Methyl-*N*-(α-methyl-3,4-methylenedioxyphenethyl)hydroxylamine
O-Methyl-*N*-(α-methyl-3,4-methylenedioxyphenethyl)hydroxylamine
α-Methyl-4-(methylthio)phenethylamine
β,3,4,5-Tetramethoxyphenethylamine
β,2,5-Trimethoxy-4-methylphenethylamine

(c) any compound (not being methoxyphenamine or a compound for the time being specified in subparagraph (a) above) structurally derived from phenethylamine, an *N*-alkylphenethylamine, α-methylphenethylamine, an *N*-alkyl-α-methylphenethylamine, α-ethylphenethylamine, or an *N*-alkyl-α-ethylphenethylamine by substitution in the ring to any extent with alkyl, alkoxy, alkylenedioxy or halide substituents, whether or not further substituted in the ring by one or more other univalent substituents. [(S.I. 1243)1977]

(d) any compound (not being a compound for the time being specified in subparagraph (a) above) structurally derived from fentanyl by modification in any of the following ways, that is to say,
  (i) by replacement of the phenyl portion of the phenethyl group by any heteromonocycle whether or not further substituted in the heterocycle;
  (ii) by substitution in the phenethyl group with alkyl, alkenyl, alkoxy, hydroxy, halogeno, haloalkyl, amino or nitro groups;
  (iii) by substitution in the piperidine ring with alkyl or alkenyl groups;
  (iv) by substitution in the aniline ring with alkyl, alkoxy, alkylenedioxy, halogeno or haloalkyl groups;
  (v) by substitution at the 4-position of the piperidine ring with any alkoxycarbonyl or alkoxyalkyl or acyloxy group;

(vi) by replacement of the *N*-propionyl group by another acyl group; [(S.I. 2230)1986]
(e) any compound (not being a compound for the time being specified in subparagraph (a) above) structurally derived from pethidine by modification in any of the following ways, that is to say,
  (i) by replacement of the 1-methyl group by an acyl, alkyl whether or not unsaturated, benzyl or phenethyl group, whether or not further substituted;
  (ii) by substitution in the piperidine ring with alkyl or alkenyl groups or with a propano bridge, whether or not further substituted;
  (iii) by substitution in the 4-phenyl ring with alkyl, alkoxy, aryloxy, halogeno or haloalkyl groups;
  (iv) by replacement of the 4-ethoxycarbonyl by any other alkoxycarbonyl or any alkoxyalkyl or acyloxy group;
  (v) by formation of an *N*-oxide or of a quaternary base. [(S.I. 2230)1986]
2. Any stereoisomeric form of a substance for the time being specified in paragraph 1 not being dextromethorphan or dextrorphan.
3. Any ester or ether of a substance for the time being specified in paragraph 1 or 2, not being a substance for the time being specified in Part II of this Schedule. [(S.I. 771)1973]
4. Any salt of a substance for the time being specified in any of paragraphs 1 to 3.
5. Any preparation or other product containing a substance or product for the time being specified in any of paragraphs 1 to 4
6. Any preparation designed for administration by injection specified in any of paragraphs 1 to 3 of Part II of this Schedule.

Table A15.2 Class B controlled drugs (Part II of Schedule 2).

1. The following substances and products, namely

(a)

Substance or product	UN Convention and Schedule	Modification Order	Schedule in Regulations
Acetyldihydrocodeine	UN1961 (II)		2
Amphetamine	UN1971 (II)		2
Codeine	UN1961 (II)		2
Dihydrocodeine	UN1961 (II)		2
Ethylmorphine (3-ethylmorphine)	UN1961 (II)		2
Glutethimide	UN1971 (III)	(S.I. 1995)1985	2
Lefetamine	UN1971 (IV)	(S.I. 1995)1985	2
Mecloqualone	UN1971 (II)	(S.I. 859)1984	2
Methaqualone	UN1971 (II)	(S.I. 859)1984	2
Methcathinone	UN1971 (I)	(S.I. 750) 1998	1
α-Methylphenethyl-hydroxylamine	Not listed	(S.I. 3932) 2001	2
Methylphenidate	UN1971 (II)		2
Methylphenobarbitone	UN1971 (IV)	(S.I. 859)1984	3
Nicocodine	UN1961 (II)		2
	UN1961 (II)	(S.I. 771)1973	2

Table A15.2 (*Continued*).

Substance or product	UN Convention and Schedule	Modification Order	Schedule in Regulations
Nicodicodine (6-nicotinoyl-dihydrocodeine)			
Norcodeine	UN1961 (II)		2
Pentazocine	UN1971 (III)	(S.I. 1995)1985	3
Phenmetrazine	UN1971 (II)		2
Pholcodine	UN1961 (II)		2
Propiram	UN1961 (II)	(S.I. 771)1973	2
Zipeprol	UN1971 (II)	(S.I. 750) 1998	2

[This note is not part of the published Schedule, but it was announced on 7 May 2008 that cannabis, cannabis resin cannabinol and cannabinol derivatives were to be re-classified to Class B (*i.e.* Part II of Schedule 2).]

(b)

Substance or product	UN 1961 or 1971 Conventions	Modification Order	Schedule in Regulations
any 5,5-disubstituted barbituric acid	UN1971 (III or IV) (Quinal-barbitone is in Schedule II)	(S.I. 859)1984	3 (Quinal-barbitone is in Schedule 2)

2. Any stereoisomeric form of a substance for the time being specified in paragraph 1 of this Part of this Schedule.
3. Any salt of a substance for the time being specified in paragraph 1 or 2 of this Part of this Schedule.
4. Any preparation or other product containing a substance or product for the time being specified in any of paragraphs 1 to 3 of this Part of this Schedule, not being a preparation falling within paragraph 6 of Part I of this Schedule.

Table A15.3 Class C controlled drugs (Part III of Schedule 2).

1. The following substances and products, namely

(a)

Substance or product	UN Convention and Schedule	Modification Order	Schedule in Regulations
Alprazolam	UN1971 (IV)	(S.I. 1995)1985	4 (Part I)
Aminorex	UN1971 (IV)	(S.I. 750)1998	4 (Part I)
Benzphetamine	UN1971 (IV)		3
Bromazepam	UN1971 (IV)	(S.I. 1995)1985	4 (Part I)
Brotizolam	UN1971 (IV)	(S.I. 750)1998	4 (Part I)
Buprenorphine	UN1971 (III)	(S.I. 1340)1989	3
Camazepam	UN1971 (IV)	(S.I. 1995)1985	4 (Part I)

Table A15.3 (*Continued*).

Substance or product	UN Convention and Schedule	Modification Order	Schedule in Regulations
Cannabinol	Not listed	(S.I. 3201)2003	1
Cannabinol derivatives	UN1971 (I) (Dronabinol is in Schedule II)	(S.I. 3201)2003	1(Dronabinol is in Schedule 2)
Cannabis and cannabis resin	UN1961 (I)	(S.I. 3201)2003	1
Cathine	UN1971 (III)	(S.I. 2230)1986	3
Cathinone	UN1971 (I)	(S.I. 2230)1986	1
Chlordiazepoxide	UN1971 (IV)	(S.I. 1995)1985	4 (Part I)
Chlorphentermine	Not listed		3

[This note is not part of the published Schedule, but it was announced on 7 May 2008 that cannabis, cannabis resin, cannabinol and cannabinol derivatives were to be reclassified to Class B (*i.e.* Part II of Schedule 2).]

Clobazam	UN1971 (IV)	(S.I. 1995)1985	4 (Part I)
Clonazepam	UN1971 (IV)	(S.I. 1995)1985	4 (Part I)
Clorazepic acid	UN1971 (IV)	(S.I. 1995)1985	4 (Part I)
Clotiazepam	UN1971 (IV)	(S.I. 1995)1985	4 (Part I)
Cloxazolam	UN1971 (IV)	(S.I. 1995)1985	4 (Part I)
Delorazepam	UN1971 (IV)	(S.I. 1995)1985	4 (Part I)
Dextropropoxyphene	Not listed	(S.I. 765)1983	2
Diazepam	UN1971 (IV)	(S.I. 1995)1985	4 (Part I)
Diethylpropion	UN1971 (IV)	(S.I. 859)1984	3
Estazolam	UN1971 (IV)	(S.I. 1995)1985	4 (Part I)
Ethchlorvynol	UN1971 (IV)	(S.I. 1995)1985	3
Ethinamate	UN1971 (IV)	(S.I. 1995)1985	3
N-Ethylamphetamine	UN1971 (IV)	(S.I. 2230)1986	4 (Part I)
Ethyl loflazepate	UN1971 (IV)	(S.I. 1995)1985	4 (Part I)
Fencamfamin	UN1971 (IV)	(S.I. 2230)1986	4 (Part I)
Fenethylline	UN1971 (II)	(S.I. 2230)1986	2
Fenproporex	UN1971 (IV)	(S.I. 2230)1986	4 (Part I)
Fludiazepam	UN1971 (IV)	(S.I. 1995)1985	4 (Part I)
Flunitrazepam	UN1971 (III)	(S.I. 1995)1985	3
Flurazepam	UN1971 (IV)	(S.I. 1995)1985	4 (Part I)
Halazepam	UN1971 (IV)	(S.I. 1995)1985	4 (Part I)
Haloxazolam	UN1971 (IV)	(S.I. 1995)1985	4 (Part I)
4-Hydroxy-n-butyric acid	UN1971 (IV)	(S.I. 1243)2003	4 (Part I)
Ketamine	Not listed	(S.I. 3178)2005	4 (Part I)
Ketazolam	UN1971 (IV)	(S.I. 1995)1985	4 (Part I)
Loprazolam	UN1971 (IV)	(S.I. 1995)1985	4 (Part I)
Lorazepam	UN1971 (IV)	(S.I. 1995)1985	4 (Part I)
Lormetazepam	UN1971 (IV)	(S.I. 1995)1985	4 (Part I)
Mazindol	UN1971 (IV)	(S.I. 1995)1985	3
Medazepam	UN1971 (IV)	(S.I. 1995)1985	4 (Part I)
Mefenorex	UN1971 (IV)	(S.I. 2230)1986	4 (Part I)
Mephentermine	Not listed		3

Table A15.3 (*Continued*).

Meprobamate	UN1971 (IV)	(S.I. 1995)1985	3
Mesocarb	UN1971 (IV)	(S.I. 750) 1998	4 (Part I)
Methyprylone	UN1971 (IV)	(S.I. 1995)1985	3
Midazolam	UN1971 (IV)	(S.I. 2589)1990	3
Nimetazepam	UN1971 (IV)	(S.I. 1995)1985	4 (Part I)
Nitrazepam	UN1971 (IV)	(S.I. 1995)1985	4 (Part I)
Nordazepam	UN1971 (IV)	(S.I. 1995)1985	4 (Part I)
Oxazepam	UN1971 (IV)	(S.I. 1995)1985	4 (Part I)
Oxazolam	UN1971 (IV)	(S.I. 1995)1985	4 (Part I)
Pemoline	UN1971 (IV)	(S.I. 1340)1989	4 (Part I)
Phendimetrazine	UN1971 (IV)		3
Phentermine	UN1971 (IV)	(S.I. 1995)1985	3
Pinazepam	UN1971 (IV)	(S.I. 1995)1985	4 (Part I)
Pipradrol	UN1971 (IV)		3
Prazepam	UN1971 (IV)	(S.I. 1995)1985	4 (Part I)
Pyrovalerone	UN1971 (IV)	(S.I. 2230)1986	4 (Part I)
Temazepam	UN1971 (IV)	(S.I. 1995)1985	3
Tetrazepam	UN1971 (IV)	(S.I. 1995)1985	4 (Part I)
Triazolam	UN1971 (IV)	(S.I. 1995)1985	4 (Part I)
Zolpidem	UN1971 (IV)	(S.I. 1243)2003	4 (Part I)

(b) [This note is not part of the published Schedule, but all of the following are unlisted in UN Conventions; except where noted, all are (S.I. 1300)1996 and all are in Schedule 4 Part II of the Regulations.]

Substance or product	*Substance or product – contd.*
4-Androstene-3,17-dione (S.I. 1243)2003	Methenolone
5-Androstene-3,17-diol (S.I. 1243)2003	Methyltestosterone
Atamestane	Metribolone
Bolandiol	Mibolerone
Bolasterone	Nandrolone
Bolazine	19-Nor-4-androstene-3,17-dione (S.I. 1243)2003
Boldenone	19-Nor-5-androstene-3,17-diol (S.I. 1243)2003
Bolenol	Norboletone
Bolmantalate	Norclostebol
Calusterone	Norethandrolone
4-Chloromethandienone	Ovandrotone
Clostebol.	Oxabolone
Drostanolone	Oxandrolone
Enestebol	Oxymesterone
Epitiostanol	Oxymetholone
Ethyloestrenol	Prasterone
Fluoxymesterone	Propetandrol
Formebolone	Quinbolone
Furazabol	Roxibolone
Mebolazine	Silandrone
Mepitiostane	Stanolone
Mesabolone	Stanozolol
Mestanolone	Stenbolone

Table A15.3 (*Continued*).

Substance or product	Substance or product – contd.
Mesterolone	Testosterone
Methandienone	Thiomesterone
Methandriol	Trenbolone

   (c) (S.I. 1300) 1996: any compound (not being Trilostane or a compound for the time being specified in subparagraph (b) above) structurally derived from 17-hydroxyandrostan-3-one or from 17-hydroxyestran-3-one by modification in any of the following ways, that is to say,
     (i) by further substitution at position 17 by a methyl or ethyl group;
     (ii) by substitution to any extent at one or more of the positions 1,2,4,6,7,9,11 or 16, but at no other position;
     (iii) by unsaturation in the carbocyclic ring system to any extent, provided that there are no more than two ethylenic bonds in any one carbocyclic ring;
     (iv) by fusion of ring A with a heterocyclic system;
   (d) any substance that is an ester or ether (or, where more than one hydroxyl function is available, both an ester and an ether) of a substance specified in subparagraph (b) or described in subparagraph (c) above or of cannabinol or a cannabinol derivative;
   (e) [This note is not part of the published Schedule, but all of the following are unlisted in UN Conventions; all are (S.I. 1300)1996 and all are in Schedule 4 Part II of the Regulations.]

Chorionic gonadotrophin (HCG)
Clenbuterol
Nonhuman chorionic gonadotrophin
Somatotropin
Somatrem
Somatropin

   2. Any stereoisomeric form of a substance for the time being specified in paragraph 1 of this Part of this Schedule, not being phenylpropanolamine.
   3. Any salt of a substance for the time being specified in paragraph 1 or 2 of this Part of this Schedule.
   4. Any preparation or other product containing a substance for the time being specified in any of paragraphs 1 to 3 of this Part of this Schedule

APPENDIX 16
The Misuse of Drugs Act 1971 – Schedule 2 (Part IV)

The following is the full text of Part IV.

MEANING OF CERTAIN EXPRESSIONS USED IN THIS SCHEDULE

For the purposes of this Schedule the following expressions (which are not among those defined in Section 37 (1) of this Act) have the meanings hereby assigned to them respectively, that is to say-

"cannabinol derivatives" means the following substances, except where contained in cannabis or cannabis resin, namely tetrahydro derivatives of cannabinol and 3-alkyl homologues of cannabinol or its tetrahydro derivatives;

"coca leaf" means the leaf of any plant of the genus *Erythroxylon* from whose leaves cocaine can be extracted either directly or by chemical transformation;

"concentrate of poppy-straw" means the material produced when poppy-straw has entered into a process for the concentration of its alkaloids;

"medicinal opium" means raw opium which has undergone the process necessary to adapt it for medicinal use in accordance with the requirements of the British Pharmacopoeia, whether it is in the form of powder or is granulated or is in any other form, and whether it is or is not mixed with neutral substances;

"opium poppy" means the plant of the species *Papaver somniferum* L;

"poppy-straw" means all parts, except the seeds, of the opium poppy, after mowing;

"raw opium" includes powdered or granulated opium but does not include medicinal opium.

APPENDIX 17
Phenethylamines added to the Misuse of Drugs Act in 2001

Table A17.1 The 34 PIHKAL substances[a] and 4-MTA added to the Act as Class A drugs in 2001 (S.I. 3932).

No.	IUPAC Name in paragraph 1(ba) of Part I of Schedule 2	Synonym[b]	Acronym	Ref.	Page
25	Allyl(α-methyl-3,4-methylenedioxyphenethyl)amine	3,4-MDO-N-allylamphetamine	MDAL	#101	719
16	2-Amino-1-(2,5-dimethoxy-4-methylphenyl)ethanol	2,5-dimethoxy-β-hydroxy-4-methylPEA	BOHD	#16	498
17	2-Amino-1-(3,4-dimethoxyphenyl)ethanol	3,4-dimethoxy-β-hydroxyPEA	DME	#57	609
20	Benzyl(α-methyl-3,4-methylenedioxyphenethyl)amine	3,4-MDO-N-benzylamphetamine	MDBZ	#103	721
02	4-Bromo-β,2,5-trimethoxyphenethylamine	4-bromo-2,5,β-trimethoxyPEA	BOB	#13	490
10	N-(4-sec-Butylthio-2,5-dimethoxy-phenethyl)-hydroxylamine	2,5-dimethoxy-4-(s)-butylthio-N-hydroxyPEA	HOT-17	#89	685
26	Cyclopropylmethyl(α-methyl-3,4-methylenedioxyphenethyl)amine	3,4-MDO-N-cyclopropylmethyl-amphetamine	MDCPM	#104	724
14	2-(4,7-Dimethoxy-2,3-dihydro-1H-indan-5-yl)-ethylamine	2,5-dimethoxy-3,4-(trimethylene)PEA	2C-G-3	#28	526
15	2-(4,7-Dimethoxy-2,3-dihydro-1H-indan-5-yl)-1-methylethylamine	2,5-dimethoxy-3,4-(trimethylene)amphetamine	G-3	#82	674
05	2-(2,5-Dimethoxy-4-methylphenyl)cyclopropylamine	2-(2,5-dimethoxy-4-methylphenyl)cyclopropylamine	DMCPA	#56	607
27	2-(1,4-Dimethoxy-2-naphthyl)ethylamine	1,4-dimethoxynaphthyl-2-ethylamine	2C-G-N	#31	535
28	2-(1,4-Dimethoxy-2-naphthyl)-1-methylethylamine	1,4-dimethoxynaphthyl-2-isopropylamine	G-N	#86	681
09	N-(2,5-Dimethoxy-4-propylthio-phenethyl)-hydroxylamine	2,5-dimethoxy-N-hydroxy-4-propylthioPEA	HOT-7	#88	683
29	2-(1,4-Dimethoxy-5,6,7,8-tetrahydro-2-naphthyl)-ethylamine	2,5-dimethoxy-3,4-(tetramethylene)PEA	2C-G-4	#29	529
30	2-(1,4-Dimethoxy-5,6,7,8-tetrahydro-2-naphthyl)-1-methylethylamine	2,5-dimethoxy-3,4-(tetramethylene)amphetamine	G-4	#83	676

Phenethylamines added to the Misuse of Drugs Act in 2001

07	α,α-Dimethyl-3,4-methylenedioxyphenethylamine	3,4-MDOphentermine	MDPH	#116	748
06	α,α-Dimethyl-3,4-methylenedioxyphenethyl(methyl)amine	3,4-MDOmephentermine	MDMP	#113	743
19	Dimethyl(α-methyl-3,4-methylenedioxyphenethyl)amine	3,4-MDO-N,N-dimethylamphetamine	MDDM	#105	725
08	N-(4-Ethylthio-2,5-dimethoxyphenethyl)-hydroxylamine	2,5-dimethoxy-4-ethylthio-N-hydroxyPEA	HOT-2	#87	682
18	4-Iodo-2,5-dimethoxy-α-methylphenethyl(dimethyl)amine	2,5-dimethoxy-N,N-dimethyl-4-iodoamphetamine	IDNNA	#90	687
12	2-(1,4-Methano-5,8-dimethoxy-1,2,3,4-tetrahydro-6-naphthyl)ethylamine	3,6-dimethoxy-4-(2-aminoethyl)benzonorbornane	2C-G-5	#30	532
13	2-(1,4-Methano-5,8-dimethoxy-1,2,3,4-tetrahydro-6-naphthyl)-1-methylethylamine	3,6-dimethoxy-4-(2-aminopropyl)benzonorbornane	G-5	#84	676
32	2-(5-Methoxy-2,2-dimethyl-2,3-dihydro-benzo[b]furan-6-yl)-1-methylethylamine	6-(2-aminopropyl)-2,2-dimethyl-5-methoxy-2,3-dihydrobenzofuran	F-22	#80	667
24	2-Methoxyethyl(α-methyl-3,4-methylenedioxy-phenethyl)amine	3,4-MDO-N-(2-methoxyethyl)-amphetamine	MDMEOET	#112	742
31	2-(5-Methoxy-2-methyl-2,3-dihydrobenzo[b]furan-6-yl)-1-methylethylamine	6-(2-aminopropyl)-5-methoxy-2-methyl-2,3-dihydrobenzofuran	F-2	#79	664
04	β-Methoxy-3,4-methylenedioxyphenethylamine	β-methoxy-3,4-MDPEA	BOH	#15	496
34	1-(3,4-Methylenedioxybenzyl)butyl(ethyl)amine	2-ethylamino-1-(3,4-MDOphenyl)-pentane	ETHYL-K	#78	663
33	1-(3,4-Methylenedioxybenzyl)butyl(methyl)amine	2-methylamino-1-(3,4-MDOphenyl)-pentane	METHYL-K	#129	781

Table A17.1 (Continued).

No.	IUPAC Name in paragraph 1(ba) of Part I of Schedule 2	Synonym[b]	Acronym	Ref.	Page
22	2-(α-Methyl-3,4-methylenedioxyphenethylamino)-ethanol	3,4-MDO-N-(2-hydroxyethyl)-amphetamine	MDHOET	#107	731
21	α-Methyl-3,4-methylenedioxyphenethyl(prop-2-ynyl)amine	3,4-MDO-N-propargyl-amphetamine	MDPL	#117	752
11	N-Methyl-N-(α-methyl-3,4-methylenedioxyphenethyl)hydroxylamine	N-hydroxy-N-methyl-3,4-MDA	FLEA	#81	671
23	O-Methyl-N-(α-methyl-3,4-methylenedioxyphenethyl)hydroxylamine	3,4-MDO-N-methoxy-amphetamine	MDMEO	#111	741
35	α-Methyl-4-(methylthio)phenethylamine	4-methylthioamphetamine	4-MTA	-	-
03	β,3,4,5-Tetramethoxyphenethylamine	3,4,5,β-tetramethoxyPEA	BOM	#17	500
01	β,2,5-Trimethoxy-4-methylphenethylamine	4-methyl-2,5,β-trimethoxyPEA	BOD	#14	492

[a] The number in the first column refers to Structure numbers A19.1 to A19.35, respectively, shown in Appendix 19. Acronym (except 4-MTA), Ref. and Page refer, respectively, to the code name, monograph number and page in PIHKAL.

[b] MDO is here, and in Table A20.1, an abbreviation for methylenedioxy

APPENDIX 18
Structural Classification of the Phenethylamines added to the Misuse of Drugs Act in 2001

Table A18.1 Structural classification[a] of the 34 PIHKAL substances and 4-MTA.

Structural Group	Substitution pattern	Substances (see Appendix 19)
1a	}β-substitution, α,β-disubstitution or α,α-disubstitution	1,2,3,4,5,6,7
1b	}	16 and 17
2a	}N-hydroxy, N-alkenyl, N-aryl, N-hydroxyalkyl, N-cyclopropylmethyl, N-alkoxyalkyl or N,N-disubstitution	8,9,10,11
2b	}	18,19,20,21,22,23,24,25 and 26
3a	}Annulated phenethylamines	12,13,14,15
3b	}	27,28,29,30,31 and 32
4	α-substitution beyond ethyl	33 and 34
5	Ring-substitution other than by alkyl, alkoxy, alkylenedioxy or halide	35

[a]Groups 1–3 are divided into those (a) where, according to PIHKAL, positive psychoactive effects may be expected, and those (b), where either no effect was detected, the effect was unpleasant or the dose was unacceptably high

Forensic Chemistry of Substance Misuse: A Guide to Drug Control
By L.A. King
© L.A. King 2009
Published by the Royal Society of Chemistry, www.rsc.org

APPENDIX 19
Molecular Structures of the Phenethylamines added to the Misuse of Drugs Act

Structure (A19.1) $\beta,2,5$-Trimethoxy-4-methylphenethylamine

Structure (A19.2) 4-Bromo-$\beta,2,5$-trimethoxyphenethylamine

Structure (A19.3) $\beta,3,4,5$-Tetramethoxyphenethylamine

Forensic Chemistry of Substance Misuse: A Guide to Drug Control
By L.A. King
© L.A. King 2009
Published by the Royal Society of Chemistry, www.rsc.org

Molecular Structures of the Phenethylamines

Structure (A19.4) β-Methoxy-3,4-methylenedioxyphenethylamine

Structure (A19.5) 2-(2,5-Dimethoxy-4-methylphenyl)cyclopropylamine

Structure (A19.6) α,α -Dimethyl-3,4-methylenedioxyphenethyl(methyl)amine

Structure (A19.7) α,α, -Dimethyl-3,4-methylenedioxyphenethylamine

Structure (A19.8) N-(4-Ethylthio-2,5-dimethoxyphenethyl)hydroxylamine

Structure (A19.9) N-(2,5-Dimethoxy-4-propylthiophenethyl)hydroxylamine

Structure (A19.10) *N*-(4-sec-Butylthio-2,5-dimethoxyphenethyl)hydroxylamine

Structure (A19.11) *N*-Methyl-*N*-(α-methyl-3,4-methylenedioxyphenethyl)-hydroxylamine

Structure (A19.12) 2-(1,4-Methano-5,8-dimethoxy-1,2,3,4-tetrahydro-6-naphthyl)ethylamine

Structure (A19.13) 2-(1,4-Methano-5,8-dimethoxy-1,2,3,4-tetrahydro-6-naphthyl)-1-methylethylamine

Structure (A19.14) 2-(4,7-Dimethoxy-2,3-dihydro-1*H*-indan-5-yl)ethylamine

Structure (A19.15) 2-(4,7-Dimethoxy-2,3-dihydro-1*H*-indan-5-yl)-1-methylethylamine

Structure (A19.16) 2-Amino-1-(2,5-dimethoxy-4-methylphenyl)ethanol

Structure (A19.17) 2-Amino-1-(3,4-dimethoxyphenyl)ethanol

Structure (A19.18) 4-Iodo-2,5-dimethoxy-α-methylphenethyl(dimethyl)amine

Structure (A19.19) Dimethyl(α-methyl-3,4-methylenedioxyphenethyl)amine

Structure (A19.20) Benzyl(α-methyl-3,4-methylenedioxyphenethyl)amine

Structure (A19.21) α-Methyl-3,4-methylenedioxyphenethyl(prop-2-ynyl)amine

Structure (A19.22) 2-(α-Methyl-3,4-methylenedioxyphenethylamino)ethanol

Structure (A19.23) *O*-Methyl-*N*-(α-methyl-3,4-methylenedioxyphenethyl)-hydroxylamine

Structure (A19.24) 2-Methoxyethyl(α-methyl-3,4-methylenedioxy-phenethyl)amine

Structure (A19.25) Allyl(α-methyl-3,4-methylenedioxyphenethyl)amine

Structure (A19.26) Cyclopropylmethyl(α-methyl-3,4-methylenedioxyphenethyl)-amine

Structure (A19.27) 2-(1,4-Dimethoxy-2-naphthyl)ethylamine

Structure (A19.28) 2-(1,4-Dimethoxy-2-naphthyl)-1-methylethylamine

Structure (A19.29) 2-(1,4-Dimethoxy-5,6,7,8-tetrahydro-2-naphthyl)ethylamine

Structure (A19.30) 2-(1,4-Dimethoxy-5,6,7,8-tetrahydro-2-naphthyl)-1-methylethylamine

Structure (A19.31) 2-(5-Methoxy-2-methyl-2,3-dihydrobenzo[*b*]furan-6-yl)-1-methylethylamine

Structure (A19.32) 2-(5-Methoxy-2,2-dimethyl-2,3-dihydrobenzo[*b*]furan-6-yl)-1-methylethylamine

Structure (A19.33) 1-(3,4-Methylenedioxybenzyl)butyl(methyl)amine

Structure (A19.34) 1-(3,4-Methylenedioxybenzyl)butyl(ethyl)amine

Structure (A19.35) α-Methyl-4-(methylthio)phenethylamine

APPENDIX 20
Derivatives of Tryptamine

Structure (A20.1) Tryptamine showing substitution patterns

Table A20.1 Derivatives of tryptamine[a] listed in TIHKAL – see Structure (A20.1).

R'	R''	$R^{\alpha 1}$	$R^{\alpha 2}$	$R^{\beta 1}$	$R^{\beta 2}$	R^2	R^1	$R^{4\ to\ 7}$	Controlled	Acronym (Name)	Ref.	Page
n-But	n-But	H	H	H	H	H	H	H	Yes	DBT	2	393
Ethyl	Ethyl	H	H	H	H	H	H	H	Yes	DET	3	396
i-Pro	i-Pro	H	H	H	H	H	H	H	Yes	DIPT	4	403
H	H	Me	H	H	H	H	H	5-MeO	No	5-MeO-α-MT	5	406
Methyl	Methyl	H	H	H	H	H	H	H	Yes	DMT	6	412
H	H	Methyl	H	H	H	Methyl	H	H	No	2,α-DMT	7	422
Methyl	H	Methyl	H	H	H	H	H	H	No	α,N-DMT	8	423
Pro	Pro	H	H	H	H	H	H	H	Yes	DPT	9	427
Ethyl	i-Pro	H	H	H	H	H	H	H	Yes	EIPT	10	431
H	H	Ethyl	H	H	H	H	H	H	Yes	α-ET (Etryptamine)	11	433
n-But	n-But	H	H	H	H	H	H	4-HO	Yes	4-HO-DBT	15	458
Ethyl	Ethyl	H	H	H	H	H	H	4-HO	Yes	4-HO-DET	16	461
i-Pro	i-Pro	H	H	H	H	H	H	4-HO	Yes	4-HO-DIPT	17	465
Methyl	Methyl	H	H	H	H	H	H	4-HO	Yes	4-HO-DMT (Psilocin)	18	468
Methyl	Methyl	H	H	H	H	H	H	5-HO	Yes	5-HO-DMT (Bufotenine)	19	473
Pro	Pro	H	H	H	H	H	H	4-HO	Yes	4-HO-DPT	20	479
Methyl	Ethyl	H	H	H	H	H	H	4-HO	Yes	4-HO-MET	21	480
Methyl	i-Pro	H	H	H	H	H	H	4-HO	Yes	4-HO-MIPT	22	481
Methyl	Pro	H	H	H	H	H	H	4-HO	Yes	4-HO-MPT	23	484
{cyclo-butyl}	{cyclo-butyl}	H	H	H	H	H	H	4-HO	No	4-HO-pyr-T	24	486
Methyl	n-But	H	H	H	H	H	H	H	Yes	MBT	27	499
i-Pro	i-Pro	H	H	H	H	H	H	4,5-MDO	No	4,5-MDO-DIPT	28	502
i-Pro	i-Pro	H	H	H	H	H	H	5,6-MDO	No	5,6-MDO-DIPT	29	503
Methyl	Methyl	H	H	H	H	H	H	4,5-MDO	No	4,5-MDO-DMT	30	505
Methyl	Methyl	H	H	H	H	H	H	5,6-MDO	No	5,6-MDO-DMT	31	507
Methyl	i-Pro	H	H	H	H	H	H	5,6-MDO	No	5,6-MDO-MIPT	32	508
Ethyl	Ethyl	H	H	H	H	Methyl	H	H	No	2-Me-DET	33	512
Methyl	Methyl	H	H	H	H	Methyl	H	H	No	2-Me-DMT	34	514
Acetyl	H	H	H	H	H	H	H	5-MeO	No	(Melatonin)	35	516
Ethyl	Ethyl	H	H	H	H	H	H	5-MeO	Yes	5-MeO-DET	36	522

Derivatives of Tryptamine

								Acronym	Ref	Page
i-Pro	i-Pro	H	H	H	H	5-MeO	Yes	5-MeO-DIPT	37	527
Methyl	Methyl	H	H	H	H	5-MeO	Yes	5-MeO-DMT	38	531
Methyl	i-Pro	H	H	H	H	4-MeO	Yes	4-MeO-MIPT	39	538
Methyl	i-Pro	H	H	H	H	5-MeO	Yes	5-MeO-MIPT	40	541
Methyl	i-Pro	H	H	H	H	5,6-di-MeO	Yes	5,6-MeO-MIPT	41	545
Methyl	H	H	H	H	H	5-MeO	Yes	5-MeO-NMT	42	546
{cyclo-butyl}	{cyclo-butyl}	H	H	H	H	5-MeO	No	5-MeO-pyr-T	43	548
Methyl	Methyl	H	H	Methyl	H	5-MeO	No	5-MeO-TMT	45	557
Methyl	Methyl	H	H	Methyl	H	5-MeS	No	5-MeS-DMT	46	560
Methyl	i-Pro	H	H	H	H	H	Yes	MIPT	47	562
H	H	Methyl	H	H	H	H	No	α-MT	48	565
H	Ethyl	H	H	H	H	H	Yes	NET	49	570
i-Pro	H	H	H	H	H	H	Yes	NIPT	49a	571
H	Methyl	H	H	H	H	H	Yes	NMT	50	573
{cyclo-butyl}	{cyclo-butyl}	H	H	H	H	H	No	pyr-T	52	577
H	H	H	H	H	H	H	No	T (Tryptamine)	53	579
H	Methyl	Methyl	H	H	H	5-MeO	No	α,N,O-TMS	55	586

[a] Acronym, Ref. and Page refer, respectively, to the code name, monograph number and page in TIHKAL. Details of compound 49a were given in TIHKAL, but it was not identified by a unique sequence number, nor was a structure shown

Table A20.2 Tryptamines, not under international control, reported to EMCDDA since 1997 under the Early Warning System.

Acronym	TIHKAL Ref.	Controlled in UK
5-HO-MIPT	Not listed: N-isopropyl-N-methyl-5-hydroxytryptamine	Yes
5-MeO-T	Not listed: 5-methoxy-tryptamine	No
5-MeO-DALT	Not listed: N,N-diallyl-5-methoxytryptamine	No
Á-MT	Not listed: α-methyltryptamine	No
4-AcO-MIPT	Not listed: the acetyl ester of 4-HO-MIPT (#22)	Yes
4-AcO-DIPT	Not listed: the acetyl ester of 4-HO-DIPT (#17)	Yes
4-AcO-DET	Not listed: the acetyl ester of 4-HO-DET (#3)	Yes
DIPT	#4	Yes
5-MeO-α-MT	#5	No
DPT	#9	Yes
4-HO-DET	#16	Yes
4-HO-DIPT	#17	Yes
4-HO-MET	#21	Yes
4-HO-MIPT	#22	Yes
5-MeO-DET	#36	Yes
5-MeO-DIPT	#37	Yes
5-MeO-DMT	#38	Yes
5-MeO-MIPT	#40	Yes
MIPT	#47	Yes

Subject Index

References in italic type refer to figures; reference in bold type refer to tables.

2C-B (4-bromo-2,5-dimethoxyphen-ethylamine) **36**, **72**
2C-B-Fly **72**, 95, 99
2C-G-3, (2(4,7-dimethoxy-2,3-dihydro-1H-indan-5-yl)ethylamine) 71, 209, 224
2C-G-4 (2-(1,4-dimethoxy-5,6,7,8-tetrahydro-2-napthyl)ethylamine) **209**, **218**, *227*
2C-G-5 **209**, **219**, *224*
2C-G-N (2-(1,4-dimethoxy-2-napthyl)-ethylamine) **209**, **218**, *227*
2C-I 28, **72**
2C-T-2 28, **72**
2C-T-7 28, **72**
4-MTA (α-methyl-4-(methylthio)phen-ethylamine) 28, 29, 36, 40, 72, 209, 220
  harm score *134*
  structure *228*
5-IAP 106
abbreviations 3–4
aceptorphine **205**
acetic anhydride **162**
  legislation 10
acetone **162**
acetyldihydrocodeine **159**, **210**
N-acetylamphetamine 97
N-acetylanthranilic acid 10, **161**
active pharmaceutical ingredients, piperazines 101
addiction, definition, xix
adhesives
  toluene 12, 16
  use prevalence, England and Wales **2**
adulterants 118
  definition, xix
Advisory Council on the Misuse of Drugs 130, 137
  1979 review 127
  benzodiazepine assessment 1–2
  cannabis
    2005–6 review 130
    2007–8 review 135–136
    recommendations 136, **137**
  methylamphetamine 131
  risk assessments
    alcohol 13
    tobacco 13
  Scale of Drug Harm 133–135
  Select Committee on Science and Technology and 132
aerosol propellants 11
agonist, definition, xix
alcohol
  harm score 133, *134*
  legislation 12–13
  see also ethanol
ALEPH-7 **72**
alfentanil 152, **205**
alkali, definition, xix

alkaloids
  definition, xix
  see also caffeine; cocaine; methyl-
    amphetamine; opiate alkaloids;
    tryptamines
alkyl nitrites 12
  harm score *134*
  see also amyl nitrite
allobarbital **58**
allyl(α-methyl-3,4-methylenedioxy-
  phenethyl)amine (MDAL) **208**,
  218, *226*
allylprodine 67, *67*, **205**
ALPHA 98
alphacetylmethadol **205**
alphameprodine 67, **205**
alphamethadol **205**
alphaprodine 67, **205**
alprazolam **156**, **211**
aluminium foil method, xix
Amanita mushrooms **116**
amfetamine *see* amphetamine
amine, xix
amineptine 36, **36**
2-amino-1-(2,5,-dimethoxy-4-methyl-
  phenyl)ethanol (BOHD) **218**, *225*
2-amino-1-(3,4-dimethylphenyl)-
  ethanol (DME) **218**, *225*
2-aminoindan 105, *106*
aminorex 95, 124, **156**, **211**
amobarbital **58**
amphetamine **97**, 173–174, **210**
  analysis 176
  chemistry 174, **174**
  derivatives 23
  deuterated 87–88
  mode of use 176
  names 176
  pharmacology 174–175
  physical form 174
  precursors 10, **161**, 163
  purity **202**
  sentencing guidelines **171**
  stereoisomerism *53*
  street price **203**
  synthesis 175–176
  see also amphetamines

amphetamine sulfate 91
amphetamines
  harm score *134*
  N-substituted 96, **97**
  New Zealand and 120–121
  precursors, legislation 10
  use prevalence, England and Wales 2
  see also amphetamine; methylam-
    phetamine; phenethylamines
amphetaminil **97**
amyl nitrite 12
  use prevalence, England and Wales 2
anabolic steroids 55–57, **213**
  definition 55
  endogenous, xxii
  esterification 51
  ethers and esters 49
  harm score *134*
  legislative control 33, **36**, 37, 153
    proposed 147
  meat products 93
  medicinal 89–90
  novel 108, **109**
  ring-numbering system *56*
  use prevalence, England and Wales 2
analgesic xx
analogue control xx
  US 123–124
  see also generic drug control
androstene derivatives 153
androstene derivatives *see* anabolic
  steroids
4-androstene-3,17-dione **156**, **213**
5-androstene-3,17-diol **156**, **213**
anileridine **206**
anthranilic acid **162**
anti-tussive, xx
API *see* active pharmaecutical ingredient
Aramah equation 170
Argyreia nervosa see Hawaiian Baby
  Woodrose
arsenic salts 8
atamestane **156**, **213**
attention deficit hyperactivity
  disorder (ADHD) 177, 198
Australia, poisons, legislation 7–8
Ayahuasca 74

Banisteriopsis caapi see Caapi
barbital **58**
barbiturates 49, 57–59, 152
    classification 135
    harm score *134*
    United Nations Convention on Psychotropic Substances classification 25–26
barium salts 8
base purity 91
base, xx
BDB, analogues 106
benzene, legislation 16
benzethidine **206**
benzodiazepines 1–2
    harm score *134*
benzoylecgonine *62*, 182
benzphetamine **42**, **97**, **211**
benzydamine 118
benzylmethylketone *see* P2P
benzylmorphine **206**
benzyl(α-methyl-3,4-methylenedioxyphenethyl)amine (MDBZ) **209**, **218**, *225*
1-benzylpiperazine *see* BZP
benzylpiperazines **101**
betacetylmethadol **206**
betameprodine **206**
betamethadol **206**
betaprodine **206**
Betts, Leah 130
bezitramide **206**
BOB **218**
BOD **220**
BOH, (β,3,4,5-tetramethoxyphenethylamine) **209**, **219**, *222*
BOHD (2-amino-1-(2,5,-dimethoxy-4-methylphenyl)ethanol) **218**, *225*
bolandiol **156**
bolasterone **156**, **213**
bolazine **156**
boldenone **156**
bolenol **156**, **213**
bolmantalate **156**, **213**
BOM **220**
British Approved Name (BAN) 1, 17
British Pharmacopia, opium definition 81

British Pharmacopia Commission 17
brolamfetamine *see* bromo-STP
bromazepam **156**, **211**
bromo-STP 45, 151, **208**
    alternative names **19**
4-bromo-2,5-dimethoxyamphetamine 45
bromodragonfly **72**, 95, 99
4-bromo-2,5-dimethoxy-*N*-ethylphenethylamine *see* N-ethyl-2C-B
4-bromo-2,5-dimethoxy-α-methylphenethylamine *see* bromo-STP
4-bromo-2,5-dimethoxyphenethylamine *see* 2C-B
4-bromo-β-2,5-trimethoxyphenethylamine (BOB) **209**, *222*
brotizolam 113, **156**, **211**
bufotenine 73, **206**, **230**
    legislative control 33
buprenorphine 128, 152, **211**
    harm score *134*
buproprion **104**
burproprion 103
butalbital **58**
butane 11
butorphanol 110, *111*
iso-butyl nitrite 12, 16
N-(4-sec-butylthio-2,5-dimethoxyphenethyl)hydroxylamine (HOT-17) **209**, **218**, *224*
γ-butyrolactone *see* GBL
BZP (benzylpiperazine) 90, 99–100, 102, 125
    EMCDDA risk assessment **29**
    harm score *134*
    precautionary principle and 138

Caapi 74, **116**
caffeine
    as adulterant 175
    legislation 14–15
California poppy 114
californine *see* lauroscholtzine
Callaghan, James 126
calusterone **156**, **213**
camazepam **156**
cannabichromenes 61
cannabidiol 60, *61*, 164–165, 178

cannabinol 60, *61*, **140**, 178, **212**
   legislative control **33**
cannabinol derivatives 59–61, 130,
     140, 153, **212**
   definition 3, 215
   see also THC
cannabis 24, 60, 177–181, **212**
   alternative names **19**
   chemistry 178
   classification 126, 127
     2005–6 review 130
     2007–8 review 135–136
     reclassification 128–129, 129–130,
        140, 153
   definition 77
   harm score *134*
   hash oil 78–79
   high-potency 80
   medical use 79, 181
   mode of use 179–180
   names 180
   origin 179
   pharmacology 178–179
   physical form 178
   purity **202**
   seeds 78
   sentencing guidelines **171**, 172
   stated cases 169
   street price **203**
   use prevalence, England and Wales **2**
   see also cannabis resin; cannabis-
     based medicines; hash oil
cannabis cigarettes 180, 201
cannabis resin 153, 177, **212**
   alternative names **19**
   purity 201, **202**
   sentencing guidelines **171**, 172
   stated cases 169
   street price **203**
   THC concentration 60
Cannabis sativa 177, 178
cannabis-based medicines 79, 181
cannabivarins 61
carfentanil 152, **206**
   legislative control **33**
carisoprodol 117
CAS, xx

Catha edulis 15
cathine 15, 16, **212**
   stereoisomerism 54
cathinone 15, 16, **212**
cathinones 102–105
   see also khat
CBD *see* cannabidiol
CBN *see* cannabinol
chat *see* khat
Chemical Industries Association 162
chemical weapons, legislation 11
Chemical Weapons Act (1996) 11
chirality *see* stereoisomers
chloral hydrate 118
chlordiazeopoxide **156**
chlordiazepoxide **212**
chloro-MDMA **72**
4-chloromethedione **157**, **213**
m-chlorophenylpiperazine 90
chlorphentermine **33**, **212**
chorionic gonadotrophin **157**, **214**
Cigarette Lighter Refill (Safety)
     Regulations (1999) 11–12
cigarettes *see* tobacco
Clarke, Charles 130
Class A drugs
   esters and ethers 49
   list of substances 205–210
   penalties associated 32
   phenethylamines 68–72
   sentencing guidelines **171**
Class B drugs **210–211**
   penalties associated 32
   sentencing guidelines **171**
Class C drugs **211–214**
   penalties associated 32, 129, 167
Class D drugs 145
clenbuterol 153, **157**, **214**
clobazam **156**, **212**
clobenzorex **97**
clomethiazole 118
clonazepam **156**, **212**
clonitazine **206**
clorazepic acid **156**, **212**
clostebol **157**, **213**
clotiazepam 113, **156**, **212**
cloxazolam **156**, **212**

CNS stimulants
  novel 95
  see also amphetamine; cocaine; ecstacy; methylamphetamine
cobalt thiolate 183
coca leaf **206**
  definition 215
coca tea 84
cocaine 24–25, **206**
  analysis 183–184
  chemistry **174**, 181–182
  harm score *134*
  medical use 184
  Misuse of Drugs regulations and 158
  mode of use 183
  origin 182–183
  pharmacology 182
  physical form 182
  processing chemicals **162**, 183
  purity 201, **202**
  sentencing guidelines **171**
  street price **203**
  structure *62*
  synthetic 46–47
  see also coca leaf; coca tea; crack cocaine; ecgonine
cocaine base 85
cocaine powder, use prevalence, England and Wales **2**
codeine 25, 151, 159, **210**
Controlled Drugs (Drug Precursors) (Community External Trade) Regulations 2008 10
controlled substances, definition 1
Controlled Substances Act (US) 39
Controlled Substances Analogue Enforcement Act (1986) (US) 123–124
Convention Against Illicit Traffic in Narcotic Drugs and Psychotropic Substances (UN1988) 10
crack cocaine 85–86, 182
  purity 201, **202**
  stated cases 169
  street price **203**
  use prevalence, England and Wales **2**
Crime and Disorder Act (1998) 167

Criminal Justice Act (2003) 32, 167
Criminal Justice (International Co-operation) Act (1990) 10, 161–163
Criminal Justice and Police Act (2001) 167
Criminal Law Act (1977), cannabis 77–78
Customs and Excise Management Act (1979) 166
cutting agents 118
4-cyano-2-dimethylamino-4,4-diphenylbutane **160**
4cyano-2-dimethylamino-4,4-diphenylbutane **208**
4-cyano-1-methyl-4-phenylpiperidine 67, **160**, **208**
cyclizine 101
cyclobarbital **58**
cyclopropylmethyl(α-methyl-3,4-methylenedioxyphenethyl)amine (MDCPM) **209**, **218**, *227*

D2PM 111–112
Dangerous Drugs Acts 30
  1951 Act, cannabis 77
dangerous substances, legislation 16
Dangerous Substances and Preparations (Safety) Regulations (2006) 16
DBT **230**
DBZP 102
de minimis principle 168
decriminalisation, xx
delorazepam **156**, **212**
Delysid 188
dementia 115
dependence, xxi
derivatives
  definition under Misuse of Drugs Act 20–21
    dialkyl derivatives 21–22
    homologues 22–23
    structural 22
designer drugs
  European legislation 27–28
  generic controls 45–48
  see also phenethylamines; tryptamines
desomorphine 206

desoxymethyltestosterone 108
DET 73, **208**, **230**
deuteration 87–88
deuterium 87
dexamphetamine 152
dextromethorphan 55, 114
dextromoramide **206**
dextropropoxyphene 151, 159, **212**
dextrorphan 55
diacetylmorphine *see* heroin
diagnostic kits 86
dialkyl phenethylamine derivatives 21–22
diamorphine *see* heroin
diampromide **206**
diastereoisomer
  definition, xxi
  see also stereoisomers
diazepam **156**, **212**
tris(2,3-dibromopropyl)phosphate 16
diethylpropion **42**, **104**, 152, **212**
diethylthiambutene **206**
N,N-diethyltryptamine *see* DET
difenoxin 67, 151, 159, **206**
dihydrocodeine *52*, 151, 159, **210**
  low-dosage preparations 88–89
dihydroetorphine 36, 37, 153, **206**
dihydromorphine **206**
dimenoxadole **206**
dimepheptanol **206**
2(4,7-dimethoxy-2,3-dihydro-1*H*-indan-5-yl)ethylamine (2C-G-3) 71, **209**, *224*
2(4,7-dimethoxy-2,3-dihydro-1*H*-indan-5-yl)-1-methethylamine (G-3) **209**, **218**, *225*
2,5-dimethoxy-α,4-dimethyl-phenethylamine *see* STP
2,5-dimethoxy-4-methylamphetamine 45
2-(2,5-dimethoxy-4-methylphenyl)-cyclopropylamine (DMCPA) **209**, **218**, *223*
2-(1,4-dimethoxy-2-napthyl)ethylamine, (2C-G-N) **209**, **218**, *227*
2-(1,4-dimethoxy-2-napthyl)-1-methylethylamine (G-N) **209**, **218**, *227*
N-(2,5-dimethoxy-4-propylthiophenethyl)hydroxylamine (HOT-7) **209**, **218**, *223*

2-(1,4-dimethoxy-5,6,7,8-tetrahydro-2-napthyl)ethylamine (2C-G-4) **209**, **218**, *227*
2-(1,4-dimethoxy-5,6,7,8-tetrahydro-2-napthyl)-1-methethylamine (G-4) **209**, **218**, *227*
N,N-dimethylamphetamine **97**
dimethylcathinone 102, **104**
α,α-dimethyl-3,4-methylenedioxyphenethylamine (MDPH) **209**, **219**, *223*
α,α-dimethyl-3,4-methylenedioxyphenethyl(methyl)amine (MDMP) **209**, **219**, *223*
dimethyl(α-methyl-3,4-methylenedioxyphenethyl)amine (MDDM) 21–22, **209**, **219**, *225*
dimethylthiambutene **206**
N,N-dimethyltryptamine *see* DMT
dioxaphetyl butyrate **206**
diphenoxylate 67, 159, **206**
diphenyl-2-pyrrolidinylmethanol 111–112
N,N-di(2-phenylisopropyl)amine **97**
dipipanone **206**
DMCPA (2-(2,5-dimethoxy-4-methylphenyl)cyclopropylamine) **209**, **218**, *223*
DME (2-amino-1-(3,4-dimethylphenyl)-ethanol) **218**, *225*
DMHP **60**
DMT 73, 117, **208**, **230**
DOB 86
  isotopic variation 87–88
DOBmisu *see* bromo-STP
dromnabinol 79
drostanolone **157**, **213**
drotebanol 151, **206**
drug abuse 1
  definition xxi
drug classification system
  Home Office proposals for review 130
  proposed changes **143–144**
  purposes 139
  reclassified drugs **140**
  reviews 126–138
  Select Committee on Science and Technology and 132
  see also offence-dependent classification
drug content 92

Drug Futures 2025 project 115
drug misuse, definition, xxi
drug offences 2, 32, 81
  sentencing guidelines 170–172
drug precursors 160, 161–163
  legislation 10–11
    proposed 146–147
Drug Trafficking Act (1994) 166
drug use
  England and Wales **2**
  prevalence 2
Drugs Act (2005) 44
  magic mushrooms 84, 132
  plants and plant products **77**
Drugs (Prevention of Misuse) Act (1964) 7, 30–31
  structure-specific control 31
Duquenois test 180

Early Warning System 28, **232**
ecgonine 182, **206**
ecgonine derivatives 61–62, **206**
ecgonine methyl ester 182
ecstacy (group of drugs) 127–128
  classification 129–130, 135
    2008 review 136
  definition, xxi
  harm score *134*
  sentencing guidelines **171**
  street price **203**
  use prevalence, England and Wales **2**
  see also MDA; MDEA; MDMA
Ehrlich's reagent 180, 183
EMCDDA *see* European Monitoring Centre for Drugs and Drug Addiction
empathogen, xxi
enantiomer, definition, xxi
enestebol **157, 213**
Ephedra vulgaris 196
ephedrine 10, 107–108, **161**, 196
  stereoisomerism 54, 107–108
ephedrone *see* methcathinone
epitiostanol **157, 213**
ergometrine **161**, 190
  legislation 10
ergot 7

ergotamine **161**
  legislation 10
Erythrooxylon coca 181
Escscholtzia californica 114
estazolam **156, 212**
esters 50–51, **157**
ethanol, *see also* alcohol
ethchlorvynol **42**, 152, **212**
ethers 51–52, **157**
ethinamate 152, **212**
ethninamate **42**
ethyl ether **162**
ethyl loflazepate **156, 212**
ETHYL-K **219**
N-ethylamphetamine **97, 156, 212**
ethylcathinone **104**
ethylmethylthiambutene **206**
ethylmorphine *52*, 151, 159, **210**
ethyloestrenol **157**
ethylone 103
N-(4-ethylthio-2,5-dimethoxyphenyl-ethyl)hydroxylamine (HOT-2) **209, 219**, *223*
eticyclidine **206**
etonitazine **206**
etorphine **206**
etoxeradine **206**
etryptamine 73, 153, **206, 230**
Eu Action Plan on Drugs 2000–2004 28
European Directives
  2001/83/EC 89–90
  solvents 12
European Monitoring Centre for Drugs and Drug Addicion (EMCDDA)
  BZP 100
  GHB 38
  new psychoactive substances, risk assessment 29
European regulations
  drug precursors 10
  (EC) No. 237/2004 10
  (EEC) No. 111/2005 10
European Union
  legislation
    1997–2005 27–28
    2005 onwards 28–29

F-2 (2-(5-methoxy-2-methyl-2,3-dihydrobenzo[*b*]furan-6-yl)-1-methylethylamine) **209**, **219**, *228*
F-22 (2-(5-methoxy-2,2-dimethyl-2,3-dihydrobenzo[*b*]furan-6-yl)-1-methylethylamine) **209**, **219**, *228*
famprofazone **97**
fencamfamin **34**, 151, **156**, **212**
fencamine **97**
fenethylline **212**
fenfluramine 71
fenproporex **97**, **156**, **212**
fentanyl *64*, **206**
fentanyls 46, 63–64
fethylline **97**
field tests 199–200
  see also Marquis field test
FLEA **220**
fludiazepam **156**, **212**
flunitrazepam **212**
4-fluorofentanyl 46
fluoxymesterone **157**
flurazepam **212**
Fly agaric **116**
formebolone **157**
formic acid 8
furazabol **157**
furethidine **206**
furfenorex **97**

G-3 (2(4,7-dimethoxy-2,3-dihydro-1*H*-indan-5-yl)-1-methethyl-amine) **209**, **218**, *225*
G-4 **209**, **218**, *227*
G-5 (2-(1,4-methano-5,8,-dimethoxy-1,2,3,4-tetrahydro-6-napthyl)-1-methylethylamine) **209**, **219**, *224*
G-N (2-(1,4-dimethoxy-2-napthyl)-1-methylethylamine) **209**, **218**, *227*
gamma-OH *see* GHB
gas chromatography 59–60, 187
gases, legislation 12
GBL 38, 107
generic drug control 45–48
  definition xxii
  New Zealand 119–123
  objections 47–48
  proposals 146
  see also analogue control
geranamine 112
GHB 28, **36**, 38, 38–39
  harm score *134*
  manufacture 38
  synonyms 38
glaucine 114
glues *see* adhesives
glutethimide **42**, 152, **210**, 1552
guilty plea policy 199–200

halazepam **156**, **212**
half-life
  definition, xxii
  see also pharmacology
hallucinogen, definition, xxii
halothane 12
haloxazolam **156**, **212**
harmine **116**
hash oil 78–79, 164–165, 179
Hawaiian Baby Woodrose 116
helium 12
heroin **206**
  alternative names **19**
  analysis 187
  chemistry **174**, 185
  harm score *134*
  medical use 187
  mode of use 186–187
  names 187
  origin and extraction 186
  pharmacology 185–186
  physical form 185
  precursors, legislation 10
  price **203**
  purity **202**
  sentencing guidelines **171**
  synthesis 50
  use prevalence, England and Wales **2**
HM Revenue and Customs 170
Hoffman, Albert 188
Home Affairs Select Committee 129–130
homologues 22–23
  definition, xxii

Subject Index   241

HOT-2 *N*-(4-ethylthio-2,5-dimethoxy-
  phenylethyl)hydroxylamine
  (HOT-2) **209**, **219**, *223*
HOT-7 (*N*-(2,5-dimethoxy-4-propyl-
  thiophenethyl)hydroxylamine)
  209, **218**, *223*
HOT-17 (*N*-(4-sec-butylthio-2,5-dimeth-
  oxyphenethyl)hydroxylamine) **209**,
  218, *224*
hydrate, xxii
hydrochloric acid **162**
hydrocodone **206**
hydrogen 87
hydromorphinol **206**
hydromorphone **206**
1-hydroxy-1-phenyl-2-aminopropane
  53–54
4-hydroxy-n-butyric acid 153, **156**, 212
5-hydroxytryptamine subtype 2
  receptor 189
17-hydroxyandrostan-3-one *56*, **157**
17-hydroxyestran-3-one *56*, **157**
N-(2-hydroxyethyl)amphetamine **97**
N-hydroxy MDA 152, **208**
  alternative names 17–18, **19**
N-hydroxy-tenamphetamine *see N*-
  hydroxy MDA
hydroxypethidine 67, **206**
γ-hydroxybutyrate *see* GHB
hypertension, xxiii
hypnotics xxii, 118

ibogaine **116**
IDNNA 21–22, **209**, **219**, *225*
impurity profiling, xxiii
indans 105–106
indenes 105–106
Independent Enquiry into the Misuse
  of Drugs Act 14–15, 127–128
injection 92–93
  heroin 186–187
INN *see* International Nonproprietary
  Name
International Nonproprietary Name
  (INN) 17, **18**
Intoxicating Substances (Supply) Act
  (1985) 11–12

4-iodo-2,5-dimethoxy-α-methylphen-
  ethyl(dimethyl)amine (IDNNA)
  21–22, **209**, **219**, *225*
ipecachuana 159
Ipomoea sp. 116
Irish Republic, generic drug control 119
isomerism *see* stereoeisomers
isomethadone **207**
isosafrole 10, **161**
isotopic variants 86–87
  case history 87–88
IUPAC, xxiii

Joint action concerning the
  information exchange, risk
  assessment and control of new
  synthetic drugs 27–28

kava kava **116**
Ketalar 39
ketamine 28, **29**, **33**, **36**, **39**, 147, 154,
  156, 212
  harm score 133–135, *134*
  use prevalence, England and Wales 2
ketazolam **156**, 212
ketobemidone **207**
khat 15–16, 135
  see also cathinones
kratom **116**

lachrymatory products 16
laughing gas (nitrous oxide) 12
lauroscholtzine 114
lefetamine **42**, 152, **210**
legalisation, xxiii
Leuckart, Rudolf 173
Leuckart synthesis xxiii
  amphetamine 175
levomethorphan **207**
levomoramide **207**
levophenacylmorphan **207**
levorphanol **207**
liquid ecstacy *see* GHB
lofentanil 63, *64*, **207**
  legislative control **33**
loprazolam **156**, 212
lorazepam **156**, 212
lormetazepam **156**, 212
low-dosage preparations 88–89

LSD 3, 65–66, 128, **207**
 analysis 190–191
 chemistry **174**, 188
 harm score *134*
 legislation 7
 mode of use 190
 names 190
 pharmacology 188–189
 physical form 188
 precursors **161**, 189–190
  legislation 10
 price **203**
 purity **202**
 sentencing guidelines **171**
 structure *187*
 synthesis 189–190
 use prevalence, England and Wales **2**
 see also lysergamide
lysergamide 31, 65, *65*, 116, **207**
 derivatives 65–66
 legislative control **33**
lysergic acid **161**
 legislation 10
lysergic acid diethylamide *see* LSD
lysergide *see* LSD

Madol 108
magic mushrooms 83–84, 132
 legislation 44
 use prevalence, England and Wales **2**
Marinol 79
Marquis field test 199–200
 amphetamine 176
 cocaine 183
 definition, xxiii
 MDMA 194
 methylamphetamine 197–198, 201
mass spectrum, xxiii
mazindol **42**, 152, **212**
MBDB 28, **29**, **72**
MBZP 125
*m*CPCPP 101, **101**
*m*CPP 101, **101**, 125
MDA 46
 abbreviation 3
 analogues 106

MDAL (allyl(α-methyl-3,4-methyl-enedioxyphenethyl)amine) 208, 218, 226
MDBZ (benzyl(α-methyl-3,4-methyl-enedioxyphenethyl)amine) 209, 218, 225
MDCPM (cyclopropylmethyl(α-methyl-3,4-methylenedioxyphenethyl)-amine) 209, 218, 227
MDDM (dimethyl(α-methyl-3,4-methyl-enedioxyphenethyl)-amine) 21–22, 209, 219, 225
MDEA 46
 alternative names **19**
MDHOET **72**, 220
MDMA 46, 68–69, 191–192
 alternative names **19**
 analysis 194
 chemistry **174**
 classification, 2008 review 136–137
 harm score *134*
 medical use 194
 other names 193
 precursors **161**
  legislation 10
 purity 201, **202**
 structure *69*
 see also ecstacy
MDMCAT *see* methylone
MDMEOET **219**
MDMP (α,α-dimethyl-3,4-methylene-dioxyphenethyl(methy)amine) **209**, **219**, *223*
MDPH (α,α-dimethyl-3,4-methylene-dioxyphenethylamine) **209**, **219**, *223*
MDPL **220**
meat products 93
mebeverine 70
mebolazine **157**
mebroqualone 122
mecloaqualone 122, 152, **210**
medazepam **156**, **212**
median purity 91
medicinal opium, definition 215

Subject Index

medicinal products 1
  definition 9
  interpretation problems 89–90
  legislation 8–9
  see also general sales list medicines; pharmacy medicines; prescription only medicines
Medicines Act (1968) 8
  alkyl nitrites 12
  medicinal product definition 9
mefenorex 97, **156**, **212**
melatonin 74
MeOPP 125
meperidine *see* pethidine
mephedrone **104**
mephentermene, legislative control **33**
mephentermine **212**
mepitiostane **157**
meprobamate 42, 117, 152, **213**
mercury salts 8
mesabolone **157**, **213**
mescaline 31, 115–116, **207**
mesocarb 96, **156**, **213**
mestanolone **157**, **213**
mesterolone **157**, **214**
metamfetamine *see* methylamphetamine
metazocine **207**
methadone **207**
  harm score *134*
methadyl acetate **207**
methandienone **157**, **214**
methandriol **157**, **214**
2-(1,4-methano-5,8,-dimethoxy-1,2,3,4-tetrahydro-6-napthyl)ethylamine (2C-G-5) **209**, **219**, *224*
2-(1,4-methano-5,8,-dimethoxy-1,2,3,4-tetrahydro-6-napthyl)-1-methylethylamine (G-5) **209**, **219**, *224*
methaqualone **210**
  classification 127, **139**
  legislation 10
  precursors **161**
methaqualone derivatives, New Zealand legislation 121–122
methcathinone 102, **104**, 153, **210**
  alternative names 19

methenolone **156**, **213**
2-(5-methoxy-2,2-dimethyl-2,3-dihydrobenzo[*b*]furan-6-yl)-1-methylethylamine (F-22) **209**, **219**, *228*
2-methoxyethyl(α-methyl-3,4-methylenedioxyphenethyl)amine (MDMEOET) **209**, **219**, *226*
β-methoxy-3,4-methylenedioxyphenethylamine (BOH) **209**, **219**, *223*
2-methoxy-4,5-methylenethiooxyamphetamine 35
methoxyphenamine 69, *70*
para-methoxyphenamine 70
methyl ethyl ketone **162**
METHYL-K (1-(3,4-methylenedioxybenzyl)butyl(methyl)amine) **209**, **219**, *228*
methyl-MA *see* PMMA
4-methyl-aminorex **208**
methylamphetamine 97, 154, **207**
  chemistry **174**, 195
  harm score *134*
  legislative review 130–131
  Marquis field test and 199–200
  mode of use 197
  other names 197
  pharmacology 195–196
  physical form 195
  precursors **161**, 196–197
  purity **202**
  synthesis 196–197
methyldesorphine **207**
methyldihydromorphine **207**
3,4-methylenedioxyamphetamine *see* MDA
1-(3,4-methylenedioxybenzyl)butyl-(ethyl)amine (ETYHL) 209, 219, 228
1-(3,4-methylenedioxybenzyl)butyl-(methyl)amine (METHYL-K) **209**, **219**, *228*
3,4-methylenedioxyethylamphetamine *see* MDEA
3,4-methylenedioxymethylamphetamine *see* MDMA

3,4-methylenedioxyphenyl-2-propanone **161**
  legislation 10
3-methylfentanyl 63, *65*
α-methylfentanyl 63, *64*
methylhexaneamine 112
methylmethaqualone 122
2-(α-methyl-3,4-methylenedioxyphenethylamino)ethanol (MDHOET) **72**, **209**, **220**, *226*
α-methyl-3,4-methylenedioxyphenethyl(prop-2-ynyl)amine (MDPL) **209**, *220*, *226*
N-methyl-*N*-(α-methyl-3,4-methylenedioxyphenethyl)hydroxylamine (FLEA) **209**, *220*, *224*
O-methyl-*N*-(α-methyl-3,4-methylenedioxyphenethyl)hydroxylamine (MDMEO) **209**, **220**, *226*
α-methyl-4-(methylthio)phenethylamine *see* 4-MTA
4-methyl-4-(methylthio)-phenethylamine *see* 4-MTA
2-methyl-3-morpholino-1,1-diphenylpropane carboxylic acid **160**
methylone 103
α-methylphenethylhydroxylamine **97**, 153
methylphenidate *134*, **210**
methylphenobarbital **58**
methylphenobarbitone **42**, **210**
α-methylphenylhydroxylamine **210**
  legislative control **33**
2-methyl-2-(phenylmethyl)-1,3-dioxolan 162
3-methyl-2-phenyloxiran 162
1-methyl-4-phenylpiperidine-4-carboxylic acid 67, **208**
1-methyl-4-phenyl-1,2,5,6-tetrahydropyridine *see* MPTP
4-methyl-α-pyrrolidinopropiophenone *see* MPPP
methyltestosterone **156**, **213**
2-(5-methoxy-2-methyl-2,3-dihydrobenzo[*b*]furan-6-yl)-1-methylethylamine (F-2) **209**, **219**, *228*
methyprylone **42**, **213**

metopon **207**
metribolone **156**, **213**
mexican sage **116**
mibolerone **156**, **213**
midazolam **213**
Misuse of Drugs Act (1971) 3, 31–32, 126–127
  additions since 2002 35–41, **36**
  Amendment Orders 32, 151–154
  cannabis 77
  changes pending 41
  classification of drugs 32
    Misuse of Drugs Regulations and **43**
    see also Class A drugs; Class B drugs; Class C drugs
  drug intermediates 160
  ecstacy 136–137
  generic controls
    anabolic steroids 55–57
    barbiturates 57–59
    cannabinols 59–61
    ecgonal derivatives 61–62
    esters 50–51
    ethers 51–52
    fentanyls 63–64
    LSD and lysergamide derivatives 65–66
    morphine pentavalent derivatives 66
    pethidines 66–67
    phenethylamines 68–72
    salts 48–49
    stereoisomers 53–55
    tryptamines 72–73
  isotopic variation and 88
  natural products 115–117
  nomenclature 17–18, **18**
    dialkyl derivatives 21–22
    redundancy 20, **21**
    'structurally derived from' 22
  offences under 32, 81
    see also sentencing guidelines
  opium 80
  PIHKAL substances 34–35
  plants and plant products **77**
  preparations for injection 92–93

Subject Index

removed substances 33–34, **34**
Schedule 2 (Part IV) 215–216
see also Independent Inquiry into the Misuse of Drugs Act
Misuse of Drugs Act (New Zealand) 119
Misuse of Drugs Amendment Act (No.2) 1987 (New Zealand) 124–125
Misuse of Drugs Regulations (2001) 3, 41–43
  diagnostic kits 86
  medicinal products 90
  Misuse of Drugs Act (1971) and **43**
  opium 81
  Schedule 4 155–157
  Schedule 5 9, 158–159
  UN1971 and **42**
Misuse of Substances Act (proposed) 141, **143–144**
Mitragyna speciosa see kratom
modafinil 112–113
modal purity 91
6-monoacetylmorphine 185
morning glory seeds 116
morpheridine **207**
morphine 159, 184–185, **207**
  esterification 50
  ethers 51–52
    see also codeine
  low-dosage preparations 89
  pentavalent derivatives 66
morphine *N*-oxide *66*, **207**
MPPP **104**
MPTP 46
muscimol **116**
myristicin **116**
myrophine **207**

N-ethyl-2C-B **72**
nabilone 79
nalbuphine 113–114
nandrolone **156, 213**
narcolepsy, xxiii
neurotransmitter, xxiv
new drugs 94–96
  see also new psychoactive substances

new psychoactive substances
  definition 28–29
  European legislation 27–28
  see also new drugs
new synthetic drugs (NSD) *see* new psychoactive substances
New Zealand 99–100
  analogue control 124–125
  emergency scheduling 125
  generic drug control 119–123
  legislation 32, 47
nicocodine **210**
nicodicodine **139**, 151, 159, **211**
nicomorphine **207**
nicotine 8, 13
  structure *14*
  see also tobacco
nimetazepam **156, 213**
nitrazepam **156, 213**
nitrous oxide 12
nootropics *see* cognitive enhancers
19-nor-4-androstene-3,17-dione **36, 156, 213**
19-nor-5-androstene-3,17-diol **36, 156, 213**
noracymethadol **207**
norboletone **156, 213**
norclostebol **156, 213**
norcodeine 151, 159, **211**
nordazepam **213**
norephedrine **161**, 175
  stereoisomerism 54
norethandrolone **156, 213**
norlevorphanol **207**
normethadone **207**
normorphine **207**
NPAS *see* new psychoactive substances
NSD *see* new psychoactive substances
Nubain *see* nalbuphine
nutmeg **116**

offence-dependent classification 142–147
offences, proposed 142–145
opiates 2
  definition, xxiv
  see also codeine; heroin; morphine; opioids; opium

opioids 113–114, 185
  definition, xxiv
  see also opiates
opium 24–25, 80–82, 159, 184–185, **207**
  definition 81–82, 215
  identification 82
  low-dosage preparations 89
  sentencing guidelines **171**
  see also codeine; morphine; poppy-straw
opium poppy 186, 215
oripavine 111
Orthoxine 69
ovandrotone 157, **213**
oxabolone **157, 213**
oxandrolone **213**
oxazepam **156, 213**
oxazolam **156, 213**
oxybate *see* GHB
oxycodone **207**
oxymesterone **157, 213**
oxymetholone **157, 213**
oxymorphone **207**

P2P *see* 1-phenyl-2-propanone
Papaver somniferum 184, 186, 216
Paracelsus 7
parahexyl **60**
paraquat 8
PCP *see* phencyclidine
pemoline 33, **34**, 95, 151, 152, **156, 213**
pentazocine 152, **211**
pentobarbital **58**
Peruvian Torch cactus 116
pesticides 8
pethidine *67*, **207**
  alternative name **19**
pethidines 46, 66–67
peyote 115–116
*p*FPP 125
pharmaceutical ingredient *see* active pharmaceutical ingredient
pharmacology
  amphetamine 174–175
  cannabis 178–179
  cocaine 182

  heroin 185–186
  LSD 188–189
  methylamphetamine 195–196
Pharmacy Act (1868) 30
Pharmacy and Poisons Act (1933) 7
pharmaecutical ingredients 8–9
phenadoxone **207**
phenazepam 113
phencyclidine (PCP) 39, 151
  precursors 146
phendimetrazine **42, 213**
phenethylamine xxiv, 35, *68*, 124
phenethylamines 18–20, 34–35, 45–46, 68–72, 99, 153, **218–220**
  dialkyl derivatives 21–22
  European legislation 28
  molecular structures 222–228
  nomenclature 23
  novel 99
  reported **72**
  structural classification 221
  see also amphetamines; cathinones; MDMA; phenylalkylamines; PIHKAL list
N-phenethyl-4-piperidone 162
phenmetrazine **211**
phenobarbital 26
phenoperidine 67, **207**
phentermine **34, 42**, 151, **213**
phenylacetic acid 10, **162**
phenylacetone *see* 1-phenyl-2-propanone
phenylalkylamine 23
phenylalkylamines 96–99
  N-substituted 96–97
  other variants 97–98
  see also phenethylamines
1-phenylethylamine, derivatives **98**
phenylpiperazines **101**
4-phenylpiperidine-4-carboxylic acid ethyl ester 67, **160, 208**
1-phenyl-2-propanone 146
1-phenyl-2-propanone (P2P) 10, **161**, 163, *163*, 175
phenylpropanolamine 54
  see also norephredrine

pholcodine *52*, 151, 159, **211**
phosphonofluoridates 11
phosphonothiolates, legislation 11
phosphoramidocyanidates, legislation 11
PIHKAL list 18–20, 34–35, 94, **218–220**
  see also phenethylamines
piminodine **207**
pinazepam **156**, **213**
piperazines 9
  generic definition 100–102
  legislative control **33**
  novel 99–102
piperidine **162**
piperonal **161**
  legislation 10
pipradrol **42**, 111, *112*, **213**
plant-based drugs, controls 46–47
PMK xxiv
PMMA 28, **29**, 70, **72**
poisons
  definition 7
  legislation 7–8
Poisons Act (1972) 8
Poisons Rules (1982) 8
Police Foundation 127
poppers 12
poppy-straw 82–83, 147, **207**
  concentrate 215
possession (of a drug) 143–145
potassium permanganate **162**
potency xxv, 92
PPP 103, **104**
prasterone **157**, **213**
prazepam **156**, **213**
precautionary principle 137–138
precursors, drug *see* drug precursors
prenylamine **97**
prescription only medicines (POM)
  ephedrine and pseudoephedrine 108
  legislation 8–9
prices 170, **203**
primary amine, xxv
α-prodines 46
proheptazine **208**
prolintane **34**, 151
properidine **208**

propetandrol **157**, **213**
propiram 151, 159, **211**
propylhexedrine **34**, 153
isopropyl nitrite 12
pseudoephedrine 107–108, **161**, 196
  legislation 10
  stereoisomerism 54, 107–108
psilocin 73, 83, 123, **208**, **230**
  esterification 50–51
Psilocybe cubensis 83
Psilocybe semilanceata 84
psilocybin 50–51, 84
  see also magic mushrooms
psychomimetic xxv
purity xxv, 91–92, 201, **202**
  value and 170
pyrovalerone 103, **104**, **156**, **213**
α-pyrrolidinopropiophenone *see* PPP

qat *see* khat
Quietlynn Ltd. 12
quinalbarbitone *see* secobarbital
quinbolone **157**, **213**

racemates xxv
  see also steroisomerism
racemoramide **208**
racemorphan **208**
racemthorphan **208**
Reach project 16
reclassification xxv, **140**
  cannabis 128–129, 129–130, **140**, 153
  see also drug classification
reduction (chemical), xxvi
reductive amination, xxvi
Reid, John 130
remifentanil **36**, 40, *41*, *64*, 153, **208**
Road Traffic Act (1972) 166
rolicyclidine **208**
roxibolone **157**, **213**
Royal Society of Arts Report (2007)
  132–133

safrole 10, **161**
salts xxvi, 48–49
  stated cases 169

Salvia divinorum **116**
Sandoz 188
Sativex 79
Scale of Drug Harm 133–135
  proposed changes 145–146
scheduled substances, definition 1
secbutabarbital **58**, 59
secobarbital 26, **58**
secondary amines, xxvi
seizures (of drugs) 2, 3, 131
Select Committee on Science and
    Technology 132
selegiline **97**
sentencing guidelines 170–172
Serious Crime Act (2007) 118
silandrone **157, 213**
Simon test xxvi
  methylamphetamines 197
smart drugs *see* cognitive enhancers
smoking 131
  cannabis 179
  methylamphetamine 196
  see also tobacco
social drugs, legislation 12–16
solvents
  harm score *134*
  legislation 11–12, 16
somatomax *see* GHB
somatotropin **157, 214**
somatrem **157, 214**
somatropin **157, 214**
stanolone **157, 213**
stanozolol **157, 213**
stated cases 168–169
stenbolone **157, 213**
stereoisomers xxvi,.53–55
  nomenclature 17–18
  stated cases 169
  THC **60**
  United Nations Convention on
      Psychotropic Drugs and 26
stimulants xxvi, 95
STP 45
  alternative names **19**
  street price 170, **203**
Sub-Coca 103

sufentanil 151, **208**
sufentanyl *64*
sulphuric acid **162**
Sweden 87–88

Tabernanthe iboga see ibogaine
tachycardia, xxvi
temazepam **42, 213**
tenocyclidine **208**
testosterone 51, *56*, **157, 214**
  meat products 93
  see also anabolic steroids
tetrahydrocannibinol *see* THC
tetrahydrocannabinol-2-oic acid 178
tetrahydrocannabinol-4-oic acid 178
tetrahydrogestinone 108
tetralins 105–106
β,3,4,5-tetramethoxyphenethylamine
    (BOH) **209, 219**, *222*
tetrazepam **156, 213**
TFMPP 125
THC xxvi, 177
  chemistry 174, **174**, 178
  esterification 51
  natural 79
  precursor substances 59–60
  variants 60
  see also cannabinols
THCA 178
thebacon **208**
thebaine **208**
theobromine 14
THG 108
thiomesterone **157, 214**
thiopentone *58*
TIHKAL 74–75, **230–232**
  lysergamide derivatives 65
tiletamine 39
tilidate 151, **208**
TMA-2 28, **72**
tobacco
  legislation 13–14
  see also nicotine
toluene 11, 12, **162**
  legislation 16
tramadol 110

Subject Index 249

tranquillisers
  use prevalence, England and Wales **2**
  see also benzodiazepines
trenbolone 157, **214**
tri(aziridin-1-yl)phosphine oxide 16
triazolam **156**, **213**
triclofos sodium 118
1-(3-trifluoromethyl-phenyl)piperazine *see* TFMPP
trilostane *57*
trimeperidine **208**
β,2,5-trimethoxy-4-methylphenethylamine (BOD) **209**, **220**, *222*
tritium 87
tryptamine xxvii
  structure *229*
tryptamine derivatives 72–73, 151, **229–232**
  European legislation 2
  New Zealand legislation 122–123
  plants containing 117
  TIHKAL drugs 65, 74–75, **230–232**
  UK legislation 46
tryptophan 72–73

UN conventions
  drugs listed 213–214
  see also United Nations Chemical Weapons convention; United Nations Convention Against Illicit Traffic in Narcotic Drugs and Psychotropic Substances; United Nations Convention on Psychotropic Substances; United Nations Single Convention on Narcotic Drugs
United Kingdom
  legislation 166–167
    future prospects 141–142
    history 30–31, 126–127
    substance control 141–142
    see also Dangerous Drugs Act; Medicines Act; Misuse of Drugs Act

United Nations Chemical Weapons Convention 11
United Nations Convention Against Illicit Traffic In Narcotic Drugs and Psychotropic Substances (1988) 26–27, 161–163
United Nations Convention on Psychotropic Substances (1971) 25–26, 32, **211–213**
  additions since 2002 **36**
  barbiturates **58**
  ethers, esters, isomers etc. 55
United Nations Single Convention on Narcotic Drugs (1961) 24–26, 30
  additions since 2002 **36**
  Schedule I substances 24–25, **205–208**
  Schedule II substances 25, **207**, **208**
  Schedule III substances 25
  Schedule IV substances 25
United States
  analogue control 123–124
  emergency scheduling provisions 125
urine *see* pharmacology
usability 168

Vatalar 39
vinylbarbital **58**

websites 204
World Anti-Doping Agency 108
World Health Organisation, substances under review 108–109
World Health Organisation Expert Committee on Drug Dependence (ECDD) 13
  zopiclone 109–110
wrap sizes **203**

Xyrem 38

zilpaterol 108
zipeprol 153, **211**
zolpidem **36**, 41, 153, **156**, **213**
zopiclone 109–110